THE BASICS OF BIOLOGY

THE BASICS OF BIOLOGY

DEEPIKA ARORA

ANMOL PUBLICATIONS PVT. LTD.
NEW DELHI - 110 002 (INDIA)

ANMOL PUBLICATIONS PVT. LTD.

H.O.: 4374/4B, Ansari Road, Darya Ganj,
New Delhi-110 002 (India)
Ph.: 23278000, 23261597

B.O.: No. 1015, Ist Main Road, BSK IIIrd Stage
IIIrd Phase, IIIrd Block
Bangalore - 560 085 (India)
Visit us at: www.anmolpublications.com

The Basics of Biology

ISBN 978-81-261-3385-7

PRINTED IN INDIA

Printed at Mehra Offset Press, Delhi.

Contents

Preface

The Basics of Biology is a well-written, non-threatening introduction to biology intended for the common person. It is designed as a reference to allow the reader to understand biological articles, books and news programmes. It has also been structured in accordance with the National Science Education Standards for high school biology so it could be used to develop curriculum.

This solid introduction provides an overview of the field; history, key concepts, and principles. Up-to-date information on ecology and genetics accompany basic coverage of evolution, body systems, and the classes of living organisms. An insert of photographs complements the black-and-white drawings, charts, and graphs, containning supplemental information, such as a discussion of the new imaging techniques used in medical diagnosis.

Students here receive an introduction to the science of biology, an overview of the history and key scientists and concepts involved in the discipline, and a topical arrangement which lends to easy browsing and information retrieval. An outstanding introduction combining simplicity of access with concrete facts.

Author

Chapter 1

Introduction: The Nature of Science and Biology

WHAT IS BIOLOGY?

Biology is the study of life. In studying biology, you will frequently run across a lot of "big words" that you will need to learn. One thing that can help you understand those words, which will also aid you in everything from reading the newspaper to communicating with your doctor, is to know the Greek or Latin (or other) derivations of the wordstems which make up those words. For example, the word "biology" is made from the wordstems bios, which means "life," and -logy which means "to study" or "the study of." Is biology important in your everyday life? How? What sorts of questions about biology are or should be important to non-majors, to general people? What branches or subdisciplines within biology might have the answers to those questions? Can you think of any other branches of biology and what they study?

The word biology is derived from the greek words/bios/meaning/life/and/logos/meaning/study/and is defined as the science of life and living organisms. An organism is a living entity consisting of one cell e.g. bacteria, or several cells e.g. animals, plants and fungi. Aspects of biological science range from the study of molecular mechanisms in cells, to the classification and behaviour of organisms, how species evolve and interaction between ecosystems.

Biology is the study of living organisms. It is concerned with the characteristics, classification, and behaviours of organisms, how species come into existence, and the interactions they have with each other and with the natural environment. Biology encompasses a broad spectrum of academic fields that are often viewed as independent disciplines. However, together they address phenomena related to living organisms (biological phenomena) over a wide range of scales, from biophysics to ecology. All concepts in biology are subject to the same laws that other branches of science obey, such as the laws of thermodynamics and conservation of energy.

Biology encompasses a broad spectrum of academic fields that are often viewed as independent disciplines. However, together they address phenomena related to living organisms (biological phenomena) over a wide range of scales, from biophysics to ecology. It is concerned with the characteristics, classification, and behaviours of organisms, how species come into existence, and the interactions they have with each other and with the natural environment. There are however unifying features. All biological phenomena are subject to the same laws that other branches of science obey, such as the laws of thermodynamics and conservation of mass. Practically the continued existence of any biological organism is always dependant on acquiring sufficient energy to overcome deleterious entropic effects.

Many of the subdisciplines of biology, such as botany, zoology, and medicine are ancient. However, biology as a unified science first developed in the nineteenth century, as scientists discovered that all living things shared certain fundamental characteristics and were best studied as a whole. Today biology is one of the most prominent scientific fields. Over a million papers are published annually in a wide array of biology and medicine journals, and biology is a standard subject of instruction at schools and universities around the world.

As such a vast field, biology is divided into a number of subdisciplines. The old division by type of organism remains with botany encompassing the study of plants, zoology the study of animals, microbiology the study of microorganisms, and other similar disciplines. The field is also divided by the scale being studied: molecular biology looks at the fundamental chemistry of life; cellular biology looks a the basic building block of all life the cell; Physiology looks at the internal structure of organism; and ecology looks at how various organisms interrelate.

Biology often overlaps with other sciences; for example, biochemistry and toxicology with biology, chemistry, and medicine; biophysics with biology and physics; stratigraphy with biology and geography; astrobiology with biology and astronomy. Social sciences such as geography, philosophy, psychology and sociology can also interact with biology, for example, in administration of biological resources, developmental biology, biogeography, evolutionary psychology and ethics.

Biology is the study of living organisms and as such, it's one of the broadest subject areas. It ranges from the sub-cellular to whole organisms and environments. At Newcastle we concentrate on studying animals, plants and micro-organisms and give you the chance to maintain breadth throughout your degree.

Biological knowledge is increasing at an unprecedented rate. New techniques such as genetic modification and genome sequencing give us extraordinary new tools for understanding how life works and for using that understanding - but with our new abilities also come new responsibilities.

Biology helps us:

- Understand ourselves and the world around us;
- Increase the quality of human life - better use of natural resources, preventing diseases and developing new ways to use living organisms, for example;
- Improve the effects humans have on our environment - such as conserving natural habitats and endangered

species or reducing the effects of environmental pollution;

- Make judgements on how we should use our knowledge.

Biology is more important than ever in today's world! Biology is the study of life in all its various and rich forms. It influences our everyday experiences, from our relationships with the environment, to the foods we consume, to the way we treat diseases. In fact, functioning in the present world without biological knowledge is a distinct handicap, since biological advances are at the forefront of social change.

HISTORY OF BIOLOGY

The history of biology traces hu nan understanding of the living world from the earliest recorded history to modern times. Though the concept of biology as a single coherent field of knowledge only arose in the 19th century, the biological sciences emerged from traditions of medicine and natural history reaching back to the ancient Greeks, particularly Galen and Aristotle, respectively.

Anaximander, a Greek philosopher who lived from 611 to 546 BC, is credited with the first written work on natural science, a classical poem entitled On Nature. In this poem, he presented what may be the first written theory of evolution He said that in the beginning there was a fish-like creature with scales etc. that arose in and lived in the world ocean. As some of these advanced, they moved onto land, shed their scaly coverings, and became the first humans.

In 570 BC, Xenophanes was born. He was one of the first people to write about his observations of fossils. He thought that fossils were an indication that there was water/mud previously in an area.

Hippocrates lived from about 400 to 300 BC. One of the things for which he is remembered is his theory that the human body was composed of the four elements (earth, air, fire, water) plus four fluids or humors: sanguis or blood, produced by the heart; choler or yellow bile, produced by the liver; melancholia or black bile, produced by the spleen; and phlegma or phlegm, produced by the brain, which corresponded with these.

Galen lived from 131 to 210 AD. He was a philosopher, physician, anatomist, and is famous for his descriptions of human anatomy which were considered authoritative for the next 1000 years. He associated the four humors with Empedocles' scheme (blood or sanguis, yellow bile or choler, black bile or melancholia, and phlegma or phlegm. He and subsequent writers worked out and added to an elaborate system that included the four organs from which the four humors came, the four seasons, the four stages in human life, and several other things that came in fours.

In the Twelfth to Fourteenth Centuries, in Europe, science, culture, etc. were at their lowest, hence this time is referred to as the Dark Ages. Much of the knowledge from the Greeks was lost in European civilization, but preserved among the Arabs.

Fig. Robert Hooke

In 1665, Robert Hooke was the first person to see and name cells. He examined (dead) cork bark with a primitive microscope and saw little cubicles which he called cells (cell = room, cubicle).

Anton van Leeuwenhoek was the first person to observe sperm cells and with his very primitive microscope, thought he saw tiny body parts in the sperm. He used this as "proof" of the idea that the homunculus was in the sperm and the mother's body just served as a place for the planted seed to grow. Apparently Leeuwenhoek, himself, did not claim the sperm was the homunculus, but that sperm provided major life-giving qualities. Regnier de Graaf, on the other hand, thought he saw the homunculus in the egg and that the sperm merely triggers its growth somehow. This split the preformationists: was the miniature adult in the egg (followers of de Graaf) or the sperm (followers of Leeuwenhoek)? Additionally, Leeuwenhoek proposed that fertilization occurs when the sperm enters the egg, but this could not actually be observed for another 100 years because of the quality of microscopes which were available. Most of the subsequent arguments as to the nature of embryos were based on speculation rather than research.

In 1668, Francesco Redi did an experiment with flies and open vs. closed flasks which contained meat. His hypothesis was that rotten meat does not turn into flies. He observed these flasks to see in which one(s) maggots developed. He found that if a flask was closed with a lid so adult flies could not get in, no maggots developed on the rotting meat within. In a flask without a lid, maggots soon were seen in the meat, followed by more adult flies. The results of this experiment disproved the idea of spontaneous generation for larger organisms, but people still thought microscopic organisms like algae or bacteria could arise that way.

Karl von Linné lived from 1707 to 1778. As was the custom among scholars of that time, he Latinized his name, which became Carolus Linnaeus. He gets credit for our present-day classification scheme and the system of two-part scientific names for organisms, thus has been given the nickname, "The Father of Taxonomy." He was the first to attempt to classify organisms for their own sake (based on things like similar body structures) rather than to serve some human use. Linnaeus believed that people were supposed to try to discover and decipher the Creator's scheme, which would become evident as organisms were grouped with "relatives." He gave two-part Latin names to each organism he knew. For example: Homo sapiens is the scientific name for humans. Note that the genus name (the first name) is always capitalized, but the species name (the second one) never is. Because this is in a foreign language (Latin), when using a computer (or having something done by a printer), it is proper to italicize scientific names. To indicate this when writing or typing a scientific name, you should underline the name. Also, note the spelling of this particular scientific name. The species name, sapiens, is one of those words that ends in "s" (like scissors), and is both singular and plural. Another thing Linnaeus did was an attempt to organize all known organisms into a taxonomic hierarchy which he invented. The levels in this hierarchy, in order are:

James Hutton lived from 1726 to 1797. In 1795, he published a paper, later refined/modified by Charles Lyell (1797-1875), in which he said that land forms can be accounted for by current mechanisms; for example, a gorge was cut by the river running through it, and was not always there. From this, he drew two conclusions:

In 1798, Rev. Thomas Malthus, a sociologist, published a paper in which he looked at conditions in the poor neighbourhoods of London. He said that in humans, the problems of disease, suffering, starvation etc. were a consequence of the potential for the human population to grow faster than technology could keep up with. Things like the supplies of food, medical care, etc. were limited in comparison to the size of the population, thus there was competition for available resources and only the strong and healthy would survive. He was, thus, the first to talk about survival of the fittest.

In either 1802 or 1809, Jean Baptiste Lamarck published his theory of evolution. His main points were:

evolution or change within a species is driven by an innate, inner striving toward greater perfection, use or disuse of various organs made them larger or smaller, accordingly, and these acquired traits could be inherited or passed on to offspring (inheritance of acquired traits).

A number of subsequent attempts were made to support or disprove this theory without the benefit of our modern knowledge of genetics. One experiment involved amputation of mouse tails for successive generations, showing that even after twenty generations, there was no effect: baby mice were still born with tails. The Jewish practice of circumcision was also cited as opposing evidence, since obviously it had caused no long-lasting change in the population and still needed to be done to each new boy baby. Lamarck's theory seemed to make sense in the light of the then-accepted theory of pangenes coming from the body parts to make up the homunculus. The classic example he used was giraffes. He felt that giraffes' necks got longer because they stretched to reach higher leaves, and this was passed on to their babies. Another example, to make the fallacy of his theory more apparent, would be two people who developed large arm muscles because they were blacksmiths, tennis players, or weight-lifters having a baby who was born with larger than normal arm muscles.

In 1828, Karl von Baer published on the developmental stages in mammalian eggs. He was able to show that an undifferentiated, single-celled egg grows into a many-celled embryo in which the cells all have different functions. This disproved the preformation theory (which said that the preformed homunculus just gets bigger).

In 1859, Charles Darwin published The Origin of Species by Means of Natural Selection or the Preservation of Favoured Races in the Struggle for Life, more commonly known as The Origin of Species. In this landmark book, he made four main points:

a. Individuals, even siblings, in a population vary (there is variation),
b. These variations can be passed to offspring (are inherited—remember: he, too, thought this happened via pangenes),
c. (from Malthus) More offspring are produced than the environment can support, so there is competition for resources, and
d. Those individuals whose characteristics make them best suited to the environment live and reproduce and have more offspring (survival of the fittest).

In 1745 - 1748, John Needham, a Scottish clergyman and naturalist showed that microorganisms flourished in various soups that had been exposed to the air. He claimed that there was a "life force" present in the molecules of all inorganic matter, including air and the oxygen in it, that could cause spontaneous generation to occur, thus accounting for the presence of bacteria in his soups. He even briefly boiled some of his soup and poured it into "clean" flasks with cork lids, and microorganisms still grew there.

A few years later (1765 - 1767), Lazzaro Spallanzani, an Italian abbot and biologist, tried several variations on Needham's soup experiments. This initiated a heated argument between Needham and Spallanzani over sterilization (boiled broth in closed vs. open containers) as a way of refuting spontaneous generation. In 1864, Louis Pasteur claimed the prize money and published the results of an experiment he did to disproved spontaneous generation in these microscopic organisms. He used flasks of broth, as above, but did not seal any of them.

In 1865, Gregor Mendel, an Austrian monk, published a paper on genetics that earned him the nickname "the Father of Modern Genetics." One of Mendel's jobs at the monastery was to care for the garden. As he went about his chores, he noticed that some of his pea plants were tall while others were short, some had purple flowers while others had white, some had yellow seeds while others had green, and some had wrinkled seeds while others had smooth seeds.

Fig: Gregor Mendel

As Mendel raised peas, he made specific crosses between certain plants and did something very unusual for biology in those days: he counted the results. From this he developed a theory of genetics that refuted the pangene/homunculus idea and enabled people to predict the outcome of a genetic cross if the genes of the parents were known. When Mendel first published his paper, the idea of the pangenes was still so deeply held that people ignored his work or dismissed it as false. It wasn't until 1900 that a couple of botanists working on other research rediscovered his work. We will discuss Mendel's theory in more depth when we talk about genetics.

In 1870 the process of mitosis, regular cell division by which one cell divides to make two cells, was observed, and researchers noticed that chromosomes, whose function was not understood, were moving around in the cell during mitosis so that each daughter cell got an exact set of them.

In 1890 the process of meiosis, a special cell division involved in producing eggs or sperm, was observed. Again, researchers did not yet understand what chromosomes were, but they did note that as a result of meiosis, each egg or sperm cell formed had half as many chromosomes as the original cell.

In 1936, Alexander Ivanovich Oparin, a Russian scientist, published The Origins of Life, in which he described hypothetical conditions which he felt would have been necessary for life to first come into existence on early Earth. This included an atmosphere of methane, ammonia, and other gases, much volcanic activity, lightening, and warm soil and water temperatures. This hypothesis was later tested by an experiment done by Stanley Miller as a graduation student under Harold Urey in 1953. We will be discussing these ideas in more depth in Biology 106.

In 1953, James Watson, an American, and Francis Crick, an Englishman, published a paper in which they proposed a hypothetical structure for DNA which also showed how DNA could be the genetic code material and suggested a means whereby it could replicate itself. Subsequent chemical analyses of DNA have upheld their prediction.

Since that time, the Modern Synthetic Theory of Evolution was developed based on the theories of Darwin, Mendel,

Malthus, Lyell, Oparin, and many others, including more recent work in genetics, geology, paleontology, and animal behaviour.

Subdisciplines of Biology:

Anatomy	The study of body parts and their locations	ana = up tom = to cut
Botany	The study of plants	botan = grass, pasture
Cytology	The study of cells	cyto = cell
Ecology	The study of the interrelations between organisms and their environment	eco = house
Entomology	The study of insects	entomol = insect
Genetics	The study of genes and heredity	gene = origin, birth
Microbiology	The study of bacteria and other microscopic organisms	micro = small
Molecular Biology	The study of the various molecules and chemical reactions that take place in organisms	molecul = a little mass
Paleontology	The study of formerly-living organisms like fossils and dinosaurs	paleo = ancient onto = being, existing
Physiology	The study of how various body parts function	physio = nature
Zoology	The study of animals	zoo = animal

Chapter 2

The Development of Evolutionary Theory

Fig. Charles Darwin

Biology came of age as a science when Charles Darwin published "On the Origin of Species." But, the idea of evolution wasn't new to Darwin. Lamarck published a theory of evolution in 1809. Lamarck thought that species arose continually from nonliving sources. These species were initially very primitive, but increased in complexity over time due to some inherent tendency.

This type of evolution is called orthogenesis. Lamarck proposed that an organism's acclimation to the environment could be passed on to its offspring. For example, he thought proto-giraffes stretched their necks to reach higher twigs. This caused their offspring to be born with longer necks. This proposed mechanism of evolution is called the inheritance of acquired characteristics. Lamarck also believed species never went extinct, although they may change into newer forms. All three of these ideas are now known to be wrong.

Darwin's contributions include hypothesizing the pattern of common descent and proposing a mechanism for evolution – natural selection. In Darwin's theory of natural selection, new variants arise continually within populations. A small percentage of these variants cause their bearers to produce more offspring than others. These variants thrive and supplant their less productive competitors. The effect of numerous instances of selection would lead to a species being modified over time.

Darwin's theory did not accord with older theories of genetics. In Darwin's time, biologists held to the theory of blending inheritance – an offspring was an average of its parents.

If an individual had one short parent and one tall parent, it would be of medium height. And, the offspring would pass on genes for medium sized offspring. If this was the case, new genetic variations would quickly be diluted out of a population. They could not accumulate as the theory of evolution required. We now know that the idea of blending inheritance is wrong.

Darwin didn't know that the true mode of inheritance was discovered in his lifetime. Gregor Mendel, in his experiments on hybrid peas, showed that genes from a mother and father do not blend. An offspring from a short and a tall parent may be medium sized; but it carries genes for shortness and tallness. The genes remain distinct and can be passed on to subsequent generations. Mendel mailed his paper to Darwin, but Darwin never opened it.

It was a long time until Mendel's ideas were accepted. One group of biologists, called biometricians, thought Mendel's laws only held for a few traits. Most traits, they claimed, were governed by blending inheritance. Mendel studied discrete traits. Two of the traits in his famous experiments were smooth versus wrinkled coat on peas. This trait did not vary continuously. In other words, peas are either wrinkled or smooth – intermediates are not found. Biometricians considered these traits aberrations. They studied continuously varying traits like size and believed most traits showed blending inheritance.

INCORPORATING GENETICS INTO EVOLUTIONARY THEORY

The discrete genes Mendel discovered would exist at some frequency in natural populations. Biologists wondered how and if these frequencies would change. Many thought that the more common versions of genes would increase in frequency simply because they were already at high frequency.

Hardy and Weinberg independently showed that the frequency of an allele would not change over time simply due to its being rare or common. Their model had several assumptions – that all alleles reproduced at the same rate, that the population size was very large and that alleles did not change in form. Later, R. A. Fisher showed that Mendel's laws could explain continuous traits if the expression of these traits were due to the action of many genes. After this, geneticists accepted Mendel's Laws as the basic rules of genetics. From this basis, Fisher, Sewall Wright and J. B. S.. Haldane founded the field of population genetics. Population genetics is a field of biology that attempts to measure and explain the levels of genetic variation in populations.

R. A. Fisher studied the effect of natural selection on large populations. He demonstrated that even very small selective differences amongst alleles could cause appreciable changes in allele frequencies over time. He also showed that the rate of adaptive change in a population is proportional to the amount of genetic variation present. This is called Fisher's Fundamental Theorem of Natural Selection. Although it is called the fundamental theorem, it does not hold in all cases. The rate at which natural selection brings about adaptation depends on the details of how selection is working. In some rare cases, natural selection can actually cause a decline in the mean relative fitness of a population.

Sewall Wright was more concerned with drift. He stressed that large populations are often subdivided into many subpopulations. In his theory, genetic drift played a more important role compared to selection. Differentiation between subpopulations, followed by migration among them, could contribute to adaptations amongst populations. Wright also came up with the idea of the adaptive landscape – an idea that remains influential to this day. Its influence remains even though P. A. P. Moran has shown that, mathematically, adaptive landscapes don't exist as Wright envisioned them. Wright extended his results of one-locus models to a two-locus case in proposing the adaptive landscape. But, unbeknownst to him, the general conclusions of the one-locus model don't extend to the two-locus case.

J. B. S. Haldane developed many of the mathematical models of natural and artificial selection. He showed that selection and mutation could oppose each other, that deleterious mutations could remain in a population due to recurrent mutation. He also demonstrated that there was a cost to natural selection, placing a limit on the amount of adaptive substitutions a population could undergo in a given time frame.

For a long time, population genetics developed as a theoretical field. But, gathering the data needed to test the theories was nearly impossible. Prior to the advent of molecular biology, estimates of genetic variability could only be inferred from levels of morphological differences in populations. Lewontin and Hubby were the first to get a good estimate of genetic variation in a population. Using the then new technique of protein electrophoresis, they showed that 30% of the loci in a population of Drosophila pseudoobscura were polymorphic. They also showed that it was likely that they could not detect all the variation that was present. Upon finding this level of variation, the question became – was this maintained by natural selection, or simply the result of genetic drift? This level of variation was too high to be explained by balancing selection.

Motoo Kimura theorized that most variation found in populations was selectively equivalent (neutral). Multiple alleles at a locus differed in sequence, but their fitnesses were the same. Kimura's neutral theory described rates of evolution and levels of polymorphism solely in terms of mutation and genetic drift. The neutral theory did not deny that natural selection acted on natural populations; but it claimed that the majority of natural variation was transient polymorphisms of neutral alleles. Selection did not act frequently or strongly enough to influence rates of evolution or levels of polymorphism.

Initially, a wide variety of observations seemed to be consistent with the neutral theory. Eventually, however, several lines of evidence toppled it. There is less variation in natural populations than the neutral theory predicts. Also, there is too much variance in rates of substitutions in different lineages to be explained by mutation and drift alone. Finally, selection itself has been shown to have an impact on levels of nucleotide variation. Currently, there is no comprehensive mathematical theory of evolution that accurately predicts rates of evolution and levels of heterozygosity in natural populations.

EVOLUTION AMONG LINEAGES

THE PATTERN OF MACROEVOLUTION

Evolution is not progress. The popular notion that evolution can be represented as a series of improvements from simple cells, through more complex life forms, to humans (the pinnacle of evolution), can be traced to the concept of the scale of nature. This view is incorrect.

All species have descended from a common ancestor. As time went on, different lineages of organisms were modified with descent to adapt to their environments. Thus, evolution is best viewed as a branching tree or bush, with the tips of each branch representing currently living species. No living organisms today are our ancestors. Every living species is as fully modern as we are with its own unique evolutionary history. No extant species are "lower life forms," atavistic stepping stones paving the road to humanity.

A related, and common, fallacy about evolution is that humans evolved from some living species of ape. This is not the case — humans and apes share a common ancestor. Both humans and living apes are fully modern species; the ancestor we evolved from was an ape, but it is now extinct and was not the same as present day apes (or humans for that matter). If it were not for the vanity of human beings, we would be classified as an ape. Our closest relatives are, collectively, the chimpanzee and the pygmy chimp. Our next nearest relative is the gorilla.

Evidence for Common Descent and Macroevolution

Microevolution can be studied directly. Macroevolution cannot. Macroevolution is studied by examining patterns in biological populations and groups of related organisms and inferring process from pattern. Given the observation of microevolution and the knowledge that the earth is billions of years old — macroevolution could be postulated. But this extrapolation, in and of itself, does not provide a compelling explanation of the patterns of biological diversity we see today. Evidence for macroevolution, or common ancestry and modification with descent, comes from several other fields of study. These include: comparative biochemical and genetic studies, comparative developmental biology, patterns of biogeography, comparative morphology and anatomy and the fossil record.

Closely related species (as determined by morphologists) have similar gene sequences. Overall sequence similarity is not the whole story, however. The pattern of differences we see in closely related genomes is worth examining.

All living organisms use DNA as their genetic material, although some viruses use RNA. DNA is composed of strings of nucleotides. There are four different kinds of nucleotides: adenine (A), guanine (G), cytosine (C) and thymine (T). Genes are sequences of nucleotides that code for proteins. Within a gene, each block of three nucleotides is called a codon. Each codon designates an amino acid (the subunits of proteins).

The three letter code is the same for all organisms (with a few exceptions). There are 64 codons, but only 20 amino acids to code for; so, most amino acids are coded for by several codons. In many cases the first two nucleotides in the codon designate the amino acid. The third position can have any of the four nucleotides and not effect how the code is translated.

A gene, when in use, is transcribed into RNA — a nucleic acid similar to DNA. (RNA, like DNA, is made up of nucleotides although t he nucleotide uracil (U) is used in place of thymine (T).) The RNA transcribed from a gene is called messenger RNA. Messenger RNA is then translated via cellular machinery called ribosomes into a string of amino acids — a protein. Some proteins function as enzymes, catalysts that speed the chemical reactions in cells. Others are structural or involved in regulating development.

Gene sequences in closely related species are very similar. Often, the same codon specifies a given amino acid in two related species, even though alternate codons could serve functionally as well. But, some differences do exist in gene sequences. Most often, differences are in third codon positions, where changes in the DNA sequence would not disrupt the sequence of the protein.

There are other sites in the genome where nucleotide differences do not effect protein sequences. The genome of eukaryotes is loaded with 'dead genes' called pseudogenes. Pseudogenes are copies of working genes that have been inactivated by mutation. Most pseudogenes do not produce full proteins. They may be transcribed, but not translated. Or, they may be translated, but only a truncated protein is produced. Pseudogenes evolve much faster than their working counterparts. Mutations in them do not get incorporated into proteins, so they have no effect on the fitness of an organism.

Introns are sequences of DNA that interrupt a gene, but do not code for anything. The coding portions of a gene are called exons. Introns are spliced out of the messenger RNA prior to translation, so they do not contribute information needed to make the protein. They are sometimes, however, involved in regulation of the gene. Like pseudogenes, introns (in general) evolve faster than coding portions of a gene.

Nucleotide positions that can be changed without changing the sequence of a protein are called silent sites. Sites where changes result in an amino acid substitution are called replacement sites. Silent sites are expected to be more polymorphic within a population and show more differences between populations. Although both silent and replacement sites receive the same amount of mutations, natural selection only infrequently allows changes at replacement sites. Silent sites, however, are not as constrained.

Kreitman was the first demonstrate that silent sites were more variable than coding sites. Shortly after the methods of DNA sequencing were discovered, he sequenced 11 alleles of the enzyme alcohol dehydrogenase (AdH). Of the 43 polymorphic nucleotide sites he found, only one resulted in a change in the amino acid sequence of the protein.

Silent sites may not be entirely selectively neutral. Some DNA sequences are involved with regulation of genes, changes in these sites may be deleterious. Likewise, although several codons code for a single amino acid, an organism may have a preferred codon for each amino acid. This is called codon bias.

If two species shared a recent common ancestor one would expect genetic information, even information such as redundant nucleotides and the position of introns or pseudogenes, to be similar. Both species would have inherited this information from their common ancestor.

The degree of similarity in nucleotide sequence is a function of divergence time. If two populations had recently separated, few differences would have built up between them. If they separated long ago, each population would have evolved numerous differences from their common ancestor (and each other). The degree of similarity would also be a function of silent versus replacement sites. Li and Graur, in their molecular evolution text, give the rates of evolution for silent vs. replacement rates. The rates were estimated from sequence comparisons of 30 genes from humans and rodents, which diverged about 80 million years ago. Silent sites evolved at an average rate of 4.61 nucleotide substitution per 10^9 years. Replacement sites evolved much slower at an average rate of 0.85 nucleotide substitutions per 10^9 years.

Groups of related organisms are 'variations on a theme' – the same set of bones are used to construct all vertebrates. The bones of the human hand grow out of the same tissue as the bones of a bat's wing or a whale's flipper; and, they share many identifying features such as muscle insertion points and ridges. The only difference is that they are scaled differently. Evolutionary biologists say this indicates that all mammals are modified descendants of a common ancestor which had the same set of bones.

Closely related organisms share similar developmental pathways. The differences in development are most evident at the end. As organisms evolve, their developmental pathway gets modified. An alteration near the end of a developmental pathway is less likely to be deleterious than changes in early development. Changes early on may have a cascading effect. Thus most evolutionary changes in development are expected to take place at the periphery of development, or in early aspects of development that have no later repercussions. For a change in early development to be propagated, the benefit of the early alteration must outweigh the consequences to later development.

Because they have evolved this way, organisms pass through the early stages of development that their ancestors passed through up to the point of divergence. So, an organism's development mimics its ancestors although it doesn't recreate it exactly. Development of the flatfish, Pleuronectes, illustrates this point. Early on, Pleuronectes develops a tail that comes to a point. In the next developmental stage, the top lobe of the tail is larger than the bottom lobe (as in sharks). When development is complete, the upper and lower lobes are equally sized. This developmental pattern mirrors the evolutionary transitions it has undergone.

Natural selection can modify any stage of a life cycle, so some differences are seen in early development. Thus, evolution does not always recapitulate ancestral forms – butterflies did not evolve from ancestral caterpillars, for example. There are differences in the appearance of early vertebrate embryos. Amphibians rapidly form a ball of cells in early development. Birds, reptiles and mammals form a disk. The shape of the early embryo is a result of different yolk concentrations in the eggs. Birds' and reptiles' eggs are heavily yolked. Their eggs develop similarly to amphibians except the yolk has deformed the shape of the embryo. The ball is stretched out and lying atop the yolk. Mammals have no yolk, but still form a disk early. This is because they have descended from reptiles. Mammals lost their yolky eggs, but retained the early pattern of development. In all these vertebrates, the pattern of cell movements is similar despite superficial differences in appearance. In addition, all types quickly converge upon a primitive, fish-like stage within a few days. From there, development diverges.

Traces of an organism's ancestry sometimes remain even when an organism's development is complete. These are called vestigial structures. Many snakes have rudimentary pelvic bones retained from their walking ancestors. Vestigial does not mean useless, it means the structure is clearly a vestige of an structure inherited from ancestral organisms. Vestigial structures may acquire new functions. In humans, the appendix now houses some immune system cells.

Closely related organisms are usually found in close geographic proximity; this is especially true of organisms with limited dispersal opportunities. The mammalian fauna of Australia is often cited as an example of this; marsupial mammals fill most of the equivalent niches that placentals fill in other ecosystems. If all organisms descended from a common ancestor, species distribution across the planet would be a function of site of origination, potential for dispersal, distribution of suitable habitat, and time since origination. In the case of Australian mammals, their physical separation from sources of placentals means potential niches were filled by a marsupial radiation rather than a placental radiation or invasion.

Natural selection can only mold available genetically based variation. In addition, natural selection provides no mechanism for advance planning. If selection can only tinker with the available genetic variation, we should expect to see examples of jury-rigged design in living species. This is indeed the case. In lizards of the genus Cnemodophorus, females reproduce parthenogenetically. Fertility in these lizards is increased when a female mounts another female and simulates copulation. These lizards evolved from sexual lizards whose hormones were aroused by sexual behaviour. Now, although the sexual mode of reproduction has been lost, the means of getting aroused (and hence fertile) has been retained.

Fossils show hard structures of organisms less and less similar to modern organisms in progressively older rocks. In addition, patterns of biogeography apply to fossils as well as extant organisms. When combined with plate tectonics, fossils provide evidence of distributions and dispersals of ancient species. For example, South America had a very distinct marsupial mammalian fauna until the land bridge formed between North and South America. After that marsupials started disappearing and placentals took their place. This is commonly interpreted as the placentals wiping out the marsupials, but this may be an over simplification.

Transitional fossils between groups have been found. One of the most impressive transitional series is the ancient reptile to modern mammal transition. Mammals and reptiles differ in skeletal details, especially in their skulls. Reptilian jaws have four bones. The foremost is called the dentary. In mammals, the dentary bone is the only bone in the lower jaw. The other bones are part of the middle ear. Reptiles have a weak jaw and a mouthful of undifferentiated teeth. Their jaw is closed by three muscles: the external, posterior and internal adductor. Each reptile tooth is single cusped. Mammals have powerful jaws with differentiated teeth. Many of these teeth, such as the molars, are multi-cusped. The temporalis and masseter muscles, derived from the external adductor, close the mammalian jaw. Mammals have a secondary palate, a bony structure separating their nostril passages and throat, so most can swallow and breathe simultaneously. Reptiles lack this.

The evolution of these traits can be seen in a series of fossils. Procynosuchus shows an increase in size of the dentary bone and the beginnings of a palate. Thrinaxodon has a reduced number of incisors, a precursor to tooth differentiation. Cynognathus (a doglike carnivore) shows a further increase in size of the dentary bone. The other three bones are located inside the back portion of the jaw. Some teeth are multicusped and the teeth fit together tightly. Diademodon (a plant eater) shows a more advanced degree of occlusion (teeth fitting tightly). Probelesodon has developed a double joint in the jaw. The jaw could hinge off two points with the upper skull. The front hinge was probably the actual hinge while the rear hinge was an alignment guide. The forward movement of a hinge point allowed for the precursor to the modern masseter muscle to anchor further forward in the jaw. This allowed for a more powerful bite. The first true mammal was Morgonucudon, a rodent-like insectivore from the late Triassic. It had all the traits common to modern mammals. These species were not from a single, unbranched lineage. Each is an example from a group of organisms along the main line of mammalian ancestry.

The strongest evidence for macroevolution comes from the fact that suites of traits in biological entities fall into a nested pattern. For example, plants can be divided into two broad categories, non- vascular (ex. mosses) and vascular. Vascular plants can be divided into seedless (ex. ferns) and seeded. Vascular seeded plants can be divided into gymnosperms (ex. pines) and flowering plants (angiosperms). Angiosperms can be divided into monocots and dicots. Each of these types of plants have several characters that distinguish them from other plants. Traits are not mixed and matched in groups of organisms. For example, flowers are only seen in plants that carry several other characters that distinguish them as angiosperms. This is the expected pattern of common descent. All the species in a group will share traits they inherited from their common ancestor. But, each subgroup will have evolved unique traits of its own. Similarities bind groups together. Differences show how they are subdivided.

The real test of any scientific theory is its ability to generate testable predictions and, of course, have the predictions borne out. Evolution easily meets this criterion. In several of the above examples, closely related organisms share X. If wedefine it closely related as sharing X, this is an empty statement. It does however, provide a prediction. If two organisms share a similar anatomy, one would then predict that their gene sequences would be more similar than a morphologically distinct organism. This has been spectacularly borne out by the recent flood of gene sequences – the correspondence to trees drawn by morphological data is very high. The discrepancies are never too great and usually confined to cases where the pattern of relationship was debated.

MECHANISMS OF MACROEVOLUTION

The following deals with mechanisms of evolution above the species level.

SPECIATION – INCREASING BIOLOGICAL DIVERSITY

Speciation is the process of a single species becoming two or more species. Many biologists think speciation is key to understanding evolution. Some would argue that certain evolutionary phenomena apply only at speciation and macroevolutionary change cannot occur without speciation. Other biologists think major evolutionary change can occur without speciation. Changes between lineages are only an extension of the changes within each lineage. In general, paleontologists fall into the former category and geneticists in the latter.

Modes of Speciation

Biologists recognize two types of speciation: allopatric and sympatric speciation. The two differ in geographical distribution of the populations in question. Allopatric speciation is thought to be the most common form of speciation. It occurs when a population is split into two (or more) geographically isolated subdivisions that organisms cannot bridge. Eventually, the two populations' gene pools change independently until they could not interbreed even if they were brought back together. In other words, they have speciated.

Sympatric speciation occurs when two subpopulations become reproductively isolated without first becoming geographically isolated. Insects that live on a single host plant provide a model for sympatric speciation. If a group of insects switched host plants they would not breed with other members of their species still living on their former host plant. The two subpopulations could diverge and speciate. Agricultural records show that a strain of the apple maggot fly Rhagolettis pomenella began infesting apples in the 1860's. Formerly it had only infested hawthorn fruit. Feder, Chilcote and Bush have shown that two races of Rhagolettis pomenella have become behaviourally isolated. Allele frequencies at six loci (aconitase 2, malic enzyme, mannose phosphate isomerase, aspartate amino-transferase, NADH-diaphorase-2, and beta-hydroxy acid dehydrogenase) are diverging. Significant amounts of linkage disequilibrium have been found at these loci, indicating that they may all be hitchhiking on some allele under selection. Some biologists call sympatric speciation microallopatric speciation to emphasize that the subpopulations are still physically separate at an ecological level.

Biologists know little about the genetic mechanisms of speciation. Some think a series of small changes in each subdivision gradually lead to speciation. The founder effect could set the stage for relatively rapid speciation, a genetic revolution in Ernst Mayr's terms. Alan Templeton hypothesized that a few key genes could change and confer reproductive isolation. He called this a genetic transilience. Lynn Margulis thinks most speciation events are caused by changes in internal symbionts. Populations of organisms are very complicated. It is likely that there are many ways speciation can occur. Thus, all of the above ideas may be correct, each in different circumstances. Darwin's book was titled "The Origin of Species" despite the fact that he did not really address this question; over one hundred and fifty years later, how species originate is still largely a mystery.

OBSERVED SPECIATIONS

Speciation has been observed. In the plant genus Tragopogon, two new species have evolved within the past 50-60 years. They are T. mirus and T. miscellus. The new species were formed when one diploid species fertilized a different diploid species and produced a tetraploid offspring. This tetraploid offspring could not fertilize or be fertilized by either of its two parent species types. It is reproductively isolated, the definition of a species.

EXTINCTION – DECREASING BIOLOGICAL DIVERSITY

ORDINARY EXTINCTION

Extinction is the ultimate fate of all species. The reasons for extinction are numerous. A species can be competitively excluded by a closely related species, the habitat a species lives in can disappear and/or the organisms that the species exploits could come up with an unbeatable defence.

Some species enjoy a long tenure on the planet while others are short- lived. Some biologists believe species are programmed to go extinct in a manner analogous to organisms being destined to die. The majority, however, believe that if the environment stays fairly constant, a well adapted species could continue to survive indefinitely.

MASS EXTINCTION

Mass extinctions shape the overall pattern of macroevolution. If you view evolution as a branching tree, it's best to picture it as one that has been severely pruned a few times in its life. The history of life on this earth includes many episodes of mass extinction in which many groups of organisms were wiped off the face of the planet. Mass extinctions are followed by periods of radiation where new species evolve to fill the empty niches left behind. It is probable that surviving a mass extinction is largely a function of luck. Thus, contingency plays a large role in patterns of macroevolution.

The largest mass extinction came at the end of the Permian, about 250 million years ago. This coincides with the formation of Pangaea II, when all the world's continents were brought together by plate tectonics. A worldwide drop in sea level also occurred at this time.

The most well-known extinction occurred at the boundary between the Cretaceous and Tertiary Periods. This called the K/T Boundary and is dated at around 65 million years ago. This extinction eradicated the dinosaurs. The K/T event was probably caused by environmental disruption brought on by a large impact of an asteroid with the earth. Following this extinction the mammalian radiation occurred. Mammals coexisted for a long time with the dinosaurs but were confined mostly to nocturnal insectivore niches. With the eradication of the dinosaurs, mammals radiated to fill the vacant niches.

Currently, human alteration of the ecosphere is causing a global mass extinction.

Punctuated Equilibrium

The theory of punctuated equilibrium is an inference about the process of macroevolution from the pattern of species documented in the fossil record. In the fossil record, transition from one species to another is usually abrupt in most geographic locales — no transitional forms are found. In short, it appears that species remain unchanged for long stretches of time and then are quickly replaced by new species. However, if wide ranges are searched, transitional forms that bridge the gap between the two species are sometimes found in small, localized areas.

For example, in Jurassic brachiopods of the genus Kutchithyris, K. acutiplicata appears below another species, K. euryptycha. Both species were common and covered a wide geographical area. They differ enough that some have argued they should be in a different genera. In just one small locality an approximately 1.25m sedimentary layer with these fossils is found. In the narrow (10 cm) layer that separates the two species, both species are found along with transitional forms. In other localities there is a sharp transition.

Eldredge and Gould proposed that most major morphological change occurs (relatively) quickly in small peripheral population at the time of speciation. New forms will then invade the range of their ancestral species. Thus, at most locations that fossils are found, transition from one species to another will be abrupt. This abrupt change will reflect replacement by migration however, not evolution. In order to find the transitional fossils, the area of speciation must be found.

There has been considerable confusion about the theory. Some popular accounts give the impression that abrupt changes in the fossil record are due to blindingly fast evolution; this is not a part of the theory.

Punctuated equilibrium has been presented as a hierarchical theory of evolution. Proponents of punctuated equilibrium see speciation as analogous to mutation and the replacement of one species by another as analogous to natural selection. This is called species selection. Speciation adds new species to the species pool just as mutation adds new alleles to the gene pool. Species selection favours one species over another just as natural selection can favour one allele over another. Evolutionary trends within a group would be the result of selection among species, not natural selection acting within species. This is the most controversial part of the theory. Many biologists agree with the pattern of macroevolution these paleontologists posit, but believe species selection is not even theoretically likely to occur.

Critics would argue that species selection is not analogous to natural selection and therefore evolution is not hierarchical. Also, the number of species produced over time is far less than the amount of different alleles that enter gene pools over time. So, the amount of adaptive evolution produced by species selection (if it did occur) would have to be orders of magnitude less than adaptive evolution within populations by natural selection.

Tests of punctuated equilibrium have been equivocal. It has been known for a long time that rates of evolution vary over time, that is not controversial. However, phylogenetic studies conflict as to whether there is a clear association between speciation and morphological change. In addition, there are major polymorphisms within some species. For example, bluegill sunfish have two male morphs.

One is a large, long-lived, mate-protecting male; the other is a smaller, shorter-lived male who sneaks matings from females guarded by large males. The existence of within species polymorphisms demonstrates that speciation is not a requirement for major morphological change.

A BRIEF HISTORY OF LIFE

Biologists studying evolution do a variety of things: population geneticists study the process as it is occurring; systematists seek to determine relationships between species and paleontologists seek to uncover details of the unfolding of life in the past. Discerning these details is often difficult, but hypotheses can be made and tested as new evidence comes to light. This section should be viewed as the best hypothesis scientists have as to the history of the planet. The material here ranges from some issues that are fairly certain to some topics that are nothing more than informed speculation.

Life evolved in the sea. It stayed there for the majority of the history of earth.

The first replicating molecules were most likely RNA. RNA is a nucleic acid similar to DNA. In laboratory studies it has been shown that some RNA sequences have catalytic capabilities. Most importantly, certain RNA sequences act as polymerases – enzymes that form strands of RNA from its monomers. This process of self replication is the crucial step in the formation of life. This is called the RNA world hypothesis.

The common ancestor of all life probably used RNA as its genetic material. This ancestor gave rise to three major lineages of life. These are: the prokaryotes ("ordinary" bacteria), archaebacteria (thermophilic, methanogenic and halophilic bacteria) and eukaryotes. Eukaryotes include protists (single celled organisms like amoebas and diatoms and a few multicellular forms such as kelp), fungi (including mushrooms and yeast), plants and animals. Eukaryotes and archaebacteria are the two most closely related of the three. The process of translation (making protein from the instructions on a messenger RNA template) is similar in these lineages, but the organization of the genome and transcription (making messenger RNA from a DNA template) is very different in prokaryotes than in eukaryotes and archaebacteria. Scientists interpret this to mean that the common ancestor was RNA based; it gave rise to two lineages that independently formed a DNA genome and hence independently evolved mechanisms to transcribe DNA into RNA.

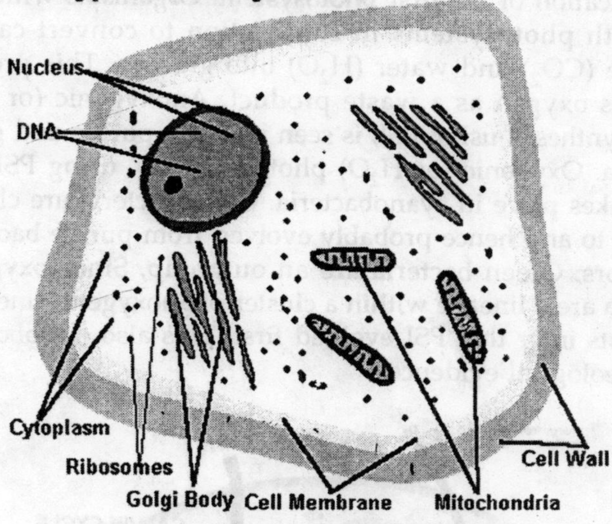

Fig. Eukaryotes

The first cells must have been anaerobic because there was no oxygen in the atmosphere. In addition, they were probably thermophilic ("heat-loving") and fermentative. Rocks as old as 3.5 billion years old have yielded prokaryotic fossils. Specifically, some rocks from Australia called the Warrawoona series give evidence of bacterial communities organized into structures called stromatolites. Fossils like these have subsequently been found all over the world. These mats of bacteria still form today in a few locales (for example, Shark Bay Australia). Bacteria are the only life forms found in the rocks for a long, long time — eukaryotes (protists) appear about 1.5 billion years ago and fungi-like things appear about 900 million years ago (0.9 billion years ago).

Photosynthesis evolved around 3.4 billion years ago. Photosynthesis is a process that allows organisms to harness sunlight to manufacture sugar from simpler precursors. The first photosystem to evolve, PSI, uses light to convert carbon dioxide (CO_2) and hydrogen sulfide (H_2S) to glucose. This process releases sulphur as a waste product. About a billion years later, a second photosystem (PS) evolved, probably from a duplication of the first photosystem. Organisms with PSII use both photosystems in conjunction to convert carbon dioxide (CO_2) and water (H_2O) into glucose. This process releases oxygen as a waste product. Anoxygenic (or H_2S) photosynthesis, using PSI, is seen in living purple and green bacteria. Oxygenic (or H_2O) photosynthesis, using PSI and PSII, takes place in cyanobacteria. Cyanobacteria are closely related to and hence probably evolved from purple bacterial ancestors. Green bacteria are an outgroup. Since oxygenic bacteria are a lineage within a cluster of anoxygenic lineages, scientists infer that PSI evolved first. This also corroborates with geological evidence.

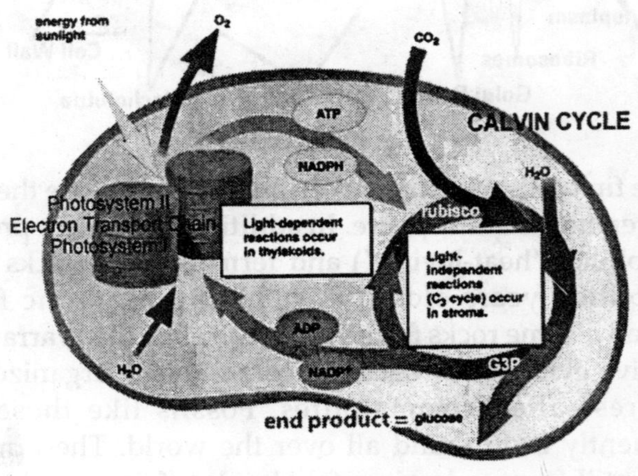

Fig. Photosynthesis

Green plants and algae also use both photosystems. In these organisms, photosynthesis occurs in organelles (membrane bound structures within the cell) called chloroplasts. These organelles originated as free living bacteria related to the cyanobacteria that were engulfed by ur-eukaryotes and eventually entered into an endosymbiotic relationship. This endosymbiotic theory of eukaryotic organelles was championed by Lynn Margulis. Originally controversial, this theory is now accepted. One key line of evidence in support of this idea came when the DNA inside chloroplasts was sequenced — the gene sequences were more similar to free-living cyanobacteria sequences than to sequences from the plants the chloroplasts resided in.

After the advent of photosystem II, oxygen levels increased. Dissolved oxygen in the oceans increased as well as atmospheric oxygen. This is sometimes called the oxygen holocaust. Oxygen is a very good electron acceptor and can be very damaging to living organisms. Many bacteria are anaerobic and die almost immediately in the presence of oxygen. Other organisms, like animals, have special ways to avoid cellular damage due to this element (and in fact require it to live.) Initially, when oxygen began building up in the environment, it was neutralized by materials already present. Iron, which existed in high concentrations in the sea was oxidized and precipitated. Evidence of this can be seen in banded iron formations from this time, layers of iron deposited on the sea floor. As one geologist put it, "the world rusted." Eventually, it grew to high enough concentrations to be dangerous to living things. In response, many species went extinct, some continued (and still continue) to thrive in anaerobic microenvironments and several lineages independently evolved oxygen respiration.

The purple bacteria evolved oxygen respiration by reversing the flow of molecules through their carbon fixing pathways and modifying their electron transport chains. Purple bacteria also enabled the eukaryotic lineage to become aerobic. Eukaryotic cells have membrane bound organelles called mitochondria that take care of respiration for the cell. These are endosymbionts like chloroplasts. Mitochondria formed this symbiotic relationship very early in eukaryotic history, all but a few groups of eukaryotic cells have mitochondria. Later, a few lineages picked up chloroplasts. Chloroplasts have multiple origins. Red algae picked up ur-chloroplasts from the cyanobacterial lineage. Green algae, the group plants evolved from, picked up different urchloroplasts from a prochlorophyte, a lineage closely related to cyanobacteria.

Animals start appearing prior to the Cambrian, about 600 million years ago. The first animals dating from just before the Cambrian were found in rocks near Adelaide, Australia. They are called the Ediacarian fauna and have subsequently been found in other locales as well. It is unclear if these forms have any surviving descendants. Some look a bit like Cnidarians (jellyfish, sea anemones and the like); others resemble annelids (earthworms). All the phyla (the second highest taxonomic category) of animals appeared around the Cambrian. The Cambrian 'explosion' may have been a result of higher oxygen concentrations enabling larger organisms with higher metabolisms to evolve. Or it might be due to the spreading of shallow seas at that time providing a variety of new niches. In any case, the radiation produced a wide variety of animals.

Some paleontologists think more animal phyla were present then than now. The animals of the Burgess shale are an example of Cambrian animal fossils. These fossils, from Canada, show a bizarre array of creatures, some which appear to have unique body plans unlike those seen in any living animals.

The extent of the Cambrian explosion is often overstated. Although quick, the Cambrian explosion is not instantaneous in geologic time. Also, there is evidence of animal life prior to the Cambrian. In addition, although all the phyla of animals came into being, these were not the modern forms we see today. Our own phylum (which we share with other mammals, reptiles, birds, amphibians and fish) was represented by a small, sliver-like thing called Pikaia. Plants were not yet present. Photosynthetic protists and algae were the bottom of the food chain. Following the Cambrian, the number of marine families leveled off at a little less than 200.

The Ordovician explosion, around 500 million years ago, followed. This 'explosion', larger than the Cambrian, introduced numerous families of the Paleozoic fauna (including crinoids, articulate brachiopods, cephalopods and corals). The Cambrian fauna, (trilobites, inarticulate brachiopods, etc.) declined slowly during this time. By the end of the Ordovician, the Cambrian fauna had mostly given way to the Paleozoic fauna and the number of marine families was just over 400. It stayed at this level until the end of the Permian period.

Plants evolved from ancient green algae over 400 million years ago. Both groups use chlorophyll a and b as photosynthetic pigments. In addition, plants and green algae are the only groups to store starch in their chloroplasts. Plants and fungi (in symbiosis) invaded the land about 400 million years ago. The first plants were moss-like and required moist environments to survive. Later, evolutionary developments such as a waxy cuticle allowed some plants to exploit more inland environments. Still mosses lack true vascular tissue to transport fluids and nutrients. This limits their size since these must diffuse through the plant. Vascular plants evolved from mosses. The first vascular land plant known is Cooksonia, a spiky, branching, leafless structure. At the same time, or shortly thereafter, arthropods followed plants onto the land. The first land animals known are myriapods – centipedes and millipedes.

Vertebrates moved onto the land by the Devonian period, about 380 million years ago. Ichthyostega, an amphibian, is the among the first known land vertebrates. It was found in Greenland and was derived from lobe-finned fishes called Rhipidistians. Amphibians gave rise to reptiles. Reptiles had evolved scales to decrease water loss and a shelled egg permitting young to be hatched on land. Among the earliest well preserved reptiles is Hylonomus, from rocks in Nova Scotia.

The Permian extinction was the largest extinction in history. It happened about 250 million years ago. The last of the Cambrian Fauna went extinct. The Paleozoic fauna took a nose dive from about 300 families to about 50. It is estimated that 96% of all species (50% of all Families) met their end. Following this event, the Modern fauna, which had been slowly expanding since the Ordovician, took over.

The Modern fauna includes fish, bivalves, gastropods and crabs. These were barely affected by the Permian extinction. The Modern fauna subsequently increased to over 600 marine families at present. The Paleozoic fauna held steady at about 100 families. A second extinction event shortly following the Permian kept animal diversity low for awhile.

During the Carboniferous (the period just prior to the Permian) and in the Permian the landscape was dominated by ferns and their relatives. After the Permian extinction, gymnosperms (ex. pines) became more abundant. Gymnosperms had evolved seeds, from seedless fern ancestors, which helped their ability to disperse. Gymnosperms also evolved pollen, encased sperm which allowed for more outcrossing.

Dinosaurs evolved from archosaur reptiles, their closest living relatives are crocodiles. One modification that may have been a key to their success was the evolution of an upright stance. Amphibians and reptiles have a splayed stance and walk with an undulating pattern because their limbs are modified from fins. Their gait is modified from the swimming movement of fish.

Splay stanced animals cannot sustain continued locomotion because they cannot breathe while they move; their undulating movement compresses their chest cavity. Thus, they must stop every few steps and breath before continuing on their way. Dinosaurs evolved an upright stance similar to the upright stance mammals independently evolved. This allowed for continual locomotion. In addition, dinosaurs evolved to be warm-blooded.

Warmbloodedness allows an increase in the vigor of movements in erect organisms. Splay stanced organisms would probably not benefit from warm- bloodedness. Birds evolved from sauriscian dinosaurs. Cladistically, birds are dinosaurs. The transitional fossil Archaeopteryx has a mixture of reptilian and avian features.

Fig. Amphibians

Angiosperms evolved from gymnosperms, their closest relatives are Gnetae. Two key adaptations allowed them to displace gymnosperms as the dominant fauna — fruits and flowers. Fruits (modified plant ovaries) allow for animal-based seed dispersal and deposition with plenty of fertilizer. Flowers evolved to facilitate animal, especially insect, based pollen dispersal. Petals are modified leaves. Angiosperms currently dominate the flora of the world — over three fourths of all living plants are angiosperms.

Insects evolved from primitive segmented arthropods. The mouth parts of insects are modified legs. Insects are closely related to annelids. Insects dominate the fauna of the world. Over half of all named species are insects. One third of this number are beetles.

The end of the Cretaceous, about 65 million years ago, is marked by a minor mass extinction. This extinction marked the demise of all the lineages of dinosaurs save the birds. Up to this point mammals were confined to nocturnal, insectivorous niches. Once the dinosaurs were out of the picture, they diversified. Morgonucudon, a contemporary of dinosaurs, is an example of one of the first mammals. Mammals evolved from therapsid reptiles. The finback reptile Diametrodon is an example of a therapsid. One of the most successful lineages of mammals is, of course, humans. Humans are neotenous apes. Neoteny is a process which leads to an organism reaching reproductive capacity in its juvenile form. The primary line of evidence for this is the similarities between young apes and adult humans. Louis Bolk compiled a list of 25 features shared between adult humans and juvenile apes, including facial morphology, high relative brain weight, absence of brow ridges and cranial crests.

The earth has been in a state of flux for 4 billion years. Across this time, the abundance of different lineages varies wildly. New lineages evolve and radiate out across the face of the planet, pushing older lineages to extinction, or relictual existence in protected refugia or suitable microhabitats. Organisms modify their environments. This can be disastrous, as in the case of the oxygen holocaust. However, environmental modification can be the impetus for further evolutionary change. Overall, diversity has increased since the beginning of life. This increase is, however, interrupted numerous times by mass extinctions. Diversity appears to have hit an all-time high just prior to the appearance of humans. As the human population has increased, biological diversity has decreased at an ever-increasing pace. The correlation is probably causal.

THE IMPORTANCE OF EVOLUTION IN BIOLOGY

"Nothing in biology makes sense except in the light of evolution." – Theodosius Dobzhansky

Evolution has been called the cornerstone of biology, and for good reasons. It is possible to do research in biology with little or no knowledge of evolution. Most biologists do. But, without evolution biology becomes a disparate set of fields. Evolutionary explanations pervade all fields in biology and brings them together under one theoretical umbrella.

We know from microevolutionary theory that natural selection should optimize the existing genetic variation in a population to maximize reproductive success. This provides a framework for interpreting a variety of biological traits and their relative importance. For example, a signal intended to attract a mate could be intercepted by predators. Natural selection has caused a trade- off between attracting mates and getting preyed upon. If you assume something other than reproductive success is optimized, many things in biology would make little sense. Without the theory of evolution, life history strategies would be poorly understood.

Macroevolutionary theory also helps explain many things about how living things work. Organisms are modified over time by cumulative natural selection. The numerous examples of jury- rigged design in nature are a direct result of this. The distribution of genetically based traits across groups is explained by splitting of lineages and the continued production of new traits by mutation. The traits are restricted to the lineages they arise in.

Details of the past also hold explanatory power in biology. Plants obtain their carbon by joining carbon dioxide gas to an organic molecule within their cells. This is called carbon fixation. The enzyme that fixes carbon is RuBP carboxlyase. Plants using C3 photosynthesis lose 1/3 to 1/2 of the carbon dioxide they originally fix. RuBP carboxlyase works well in the absence of oxygen, but poorly in its presence. This is because photosynthesis evolved when there was little gaseous oxygen present. Later, when oxygen became more abundant, the efficiency of photosynthesis decreased. Photosynthetic organisms compensated by making more of the enzyme. RuBP carboxylase is the most abundant protein on the planet partially because it is one of the least efficient.

Ecosystems, species, organisms and their genes all have long histories. A complete explanation of any biological trait must have two components. First, a proximal explanation – how does it work? And second, an ultimate explanation – what was it modified from? For centuries humans have asked, "Why are we here?" The answer to that question lies outside the realm of science. Biologists, however, can provide an elegant answer to the question, "How did we get here?"

Chapter 3

Life

Life is an attribute of objects that are considered alive. There are a vast variety of living organisms. Properties common to these organisms – plants, animals, fungi, protists, archaea and bacteria – are being carbon; water-based, and being cellular with complex organization. They undergo metabolism; possess a capacity to grow; respond to stimuli, reproduce and through natural selection adapt to their environment in succeeding generations. An entity with the above properties is considered to be an organism. However, not every definition of life considers all of these properties to be essential. For example, the capacity for descent with modification is often taken as the only essential property of life. This definition notably includes viruses, which do not qualify under narrower definitions as they are acellular and do not metabolise. Broader definitions of life may also include theoretical non-carbon-based life and other alternative biology.

There is no universal definition of life; there are a variety of definitions proposed by different scientists.

A conventional definition: Often scientists say that life is a characterstic of organisms that exhibit the following phenomena:

Homeostasis: Regulation of the internal environment to maintain a constant state; for example, sweating to reduce temperature.

Organization: Being composed of one or more cells, which are the basic units of life.

Metabolism: Consumption of energy by converting nonliving material into cellular components (anabolism) and decomposing organic matter (catabolism). Living things require energy to maintain internal organization (homeostasis) and to produce the other phenomena associated with life.

Growth: Maintenance of a higher rate of synthesis than catalysis. A growing organism increases in size in all of its parts, rather than simply accumulating matter. The particular species begins to multiply and expand as the evolution continues to flourish.

Adaptation: The ability to change over a period of time in response to the environment. This ability is fundamental to the process of evolution and is determined by the organism's heredity as well as the composition of metabolized substances, and external factors present.

Response to stimuli: A response can take many forms, from the contraction of a unicellular organism when touched to complex reactions involving all the senses of higher animals. A response is often expressed by motion, for example, the leaves of a plant turning toward the sun or an animal chasing its prey.

Reproduction: The ability to produce new organisms. Reproduction can be the division of one cell to form two new cells. Usually the term is applied to the production of a new individual (either asexually, from a single parent organism, or sexually, from at least two differing parent organisms), although strictly speaking it also describes the production of new cells in the process of growth.

However, others cite several limitations of this definition. Thus, many members of several species do not reproduce, possibly because they belong to specialized sterile castes (such as ant workers), these are still considered forms of life. One could say that the property of life is inherited; hence, sterile hybrid species are considered life although not themselves capable of reproduction. It is also worth noting that non-reproducing individuals may still help the spread of their genes through such mechanisms as kin selection.

Viruses and aberrant prion proteins are often considered replicators rather than forms of life, a distinction warranted because they cannot reproduce without very specialized substrates such as host cells or proteins, respectively. However, most forms of life rely on foods produced by other species, or at least the specific chemistry of Earth's environment.

Still others contest such definitions of life on philosophical grounds. They offer the following as examples of life: viruses which reproduce; storms or flames which "burn"; certain computer software programmes which are programmed to mutate and evolve; future software programmes which may evince (even high-order) behaviour; machines which can move; and some forms of proto-life consisting of metabolizing cells without the ability to reproduce. Still, most scientists would not call such phenomena expressive of life. Generally all seven characteristics are required for a population to be considered a life form.

The systemic definition of life is that living things are self-organizing and autopoietic (self-producing). These objects are not to be confused with dissipative structures (e.g. fire).

Variations of this definition include Stuart Kauffman's definition of life as an autonomous agent or a multi-agent system capable of reproducing itself or themselves, and of completing at least one thermodynamic work cycle.

Yet other definitions of life are:

Living things are systems that tend to respond to changes in their environment, and inside themselves, in such a way as to promote their own continuation.

Life is a charateristic of self-organizing, cannibalistic systems consisting of a population of replicators that are capable of mutation, around most of which homeostatic, metabolizing organisms evolve. This definition does not include flames, but does include worker ants, viruses and mules. Self reproduction and energy consumption is only one means for a system to promote its own continuation. This explains why bees can be alive and yet commit suicide in defending their hive. In this case the whole colony works as such a living system.

Type of organization of matter producing various interacting forms of variable complexity, whose main property is to replicate almost perfectly by using matter and energy available in their environment to which they may adapt. In this definition "almost perfectly" relates to mutations happening during replication of organisms that may have adaptative benefits.

Contradictions in Definitions

Most of the definitions are focused on differentiating between living material and non-living material. It is still not clear whether there exists computer software that can be considered as a living. Some computer software show certain resemblance with life. It is not difficult in algorithmic sense to create software which fulfills some base criteria (growing, reproduction, reaction to the changes in the environment, etc.).

Consider an artificial environment with certain properties (for example an system which is calling the processes with different input). If the process is able to solve the problem (respond with a correct output) in a given time, it can survive, otherwise it is deleted. If the process solves the problem before the other processes it gets higher part from the computer resources. This is a very simple model of an environment in which the life has been created. More sophisticated environments can be created. Although in most of the simulations the systems collapse relatively fast (for example non of the processes are able to adapt and solve the given problem) it has not yet been formally proved that such an environment can not exist permanently. Such an environment could met all the criteria of some simple life definitions.

If ever such a theorem will be formally proved, it will be necessary to reconsider the definitions or admit that the in some sense the computer world is a good approximation of the real one.

ORIGIN OF LIFE

Although it cannot be pinpointed exactly, evidence suggests that life on Earth has existed for about 3.7 billion years .There is no truly "standard" model for the origin of life, but most currently accepted scientific models build in one way or another on the following discoveries, which are listed roughly in order of postulated emergence:

Plausible pre-biotic conditions result in the creation of the basic small molecules of life. This was demonstrated in the Miller-Urey experiment.

Phospholipids spontaneously form lipid bilayers, the basic structure of a cell membrane.

Procedures for producing random RNA molecules can produce ribozymes, which are able to produce more of themselves under very specific conditions.

There are many different hypotheses regarding the path that might have been taken from simple organic molecules to protocells and metabolism. Many models fall into the "genes-first" category or the "metabolism-first" category, but a recent trend is the emergence of hybrid models that do not fit into either of these categories.

EXTRATERRESTRIAL LIFE

Earth is the only planet in the universe known to harbour life. The Drake equation has been used to estimate the probability of life elsewhere, but scientists disagree on many of the values of variables in this equation (although strictly speaking Drake equation estimates the number of extraterrestrial civilizations in our galaxy with which we might come in contact - not probability of life elsewhere). Depending on those values, the equation may either suggest that life arises frequently or infrequently.

ORIGIN OF LIFE CONCEPTS

In the physical sciences, abiogenesis, the question of the origin of life, is the study of how life on Earth might have evolved from non-life sometime between 3.9 and 3.5 billion years ago. This topic also includes theories and ideas regarding possible extra-planetary or extra-terrestrial origin of life hypotheses, thought to have possibly occurred over the last 13.7 billion years in the evolution of the known universe since the big bang.

Origin of life studies is a limited field of research despite its profound impact on biology and human understanding of the natural world. Progress in this field is generally slow and sporadic, though it still draws the attention of many due to the eminence of the question being investigated. A few facts give insight into the conditions in which life may have emerged, but the mechanisms by which non-life became life are still elusive.

Asking the question about different theories on the origin of life is similar to asking the question about the possible ways other entities like a house, a car, or a factory could have come into existence. There are only three possible ways that such things could have come into existence. They are as follows:

- Random, natural unguided forces or processes
- Designed and created by a designer
- Self generation or ability to create, inherent or designed into matter.

Arguments that life spores were transported to our planet, the earth, still requires answering the same question. Which possible theory was the correct mechanism to explain the generation of life at the remote location where the spores originated?

We all know that a car or a factory is designed and created by a designer. Nobody would question that. What we have learned is that a car is both far less complex than a single living cell and cannot reproduce itself like some living cells can. Yet some of us are willing to consider that such living cells came about by a random unguided process from inorganic matter.

A very modern complex automated factory is more similar to the complexity and function of a living cell. Although it is considerably less complex, much larger and unable to reproduce itself, at least it can produce whatever it was designed to produce. Yet nobody would believe that such a factory could come into existence by a random unguided process. Why do we believe that life could have come into existence by a random unguided process?

Even if we have the absurd belief that an extremely complex living cell could have came into existence through random processes, how much more absurd is it to believe that a cell would come into existence that could reproduce itself. None of us believe that a complex factory could come into existence by a random process. How much more absurd is it that a factory that could reproduce itself could have come into existence by random chance? Man with his great intelligence has not considered designing a factory that could reproduce itself even on a macroscopic scale. How much more intelligence is required to create life with extremely complex, "machine like" cells that can reproduce themselves on a microscopic level.

Some evolutionists speculate that matter has inherent or creative power built into it. If random chance cannot produce non-reproductive life from matter (nobody has shown that it can) how can the much, much more complex inherent creative ability of matter be credited to random chance?

Astrophysicist, Sir Fred Hoyle and his colleague Chandra Wickramasinghe argued chance processes could not have formed the biochemical machinery of the cell, especially the enzymes. In their book, "Evolution from Space," they estimated the probability of forming a single enzyme of protein at random, in the rich ocean of amino acids, was no more than 10 to the 20th power. They then calculated the likelihood of forming by chance all of the more than 2000 enzymes used in the life forms on earth. This probability was calculated at one in 10 to the 40,000th power.

A vivid analogy from Hoyle became a well-known cliché. "Belief in chemical evolution of the first cell from lifeless chemicals is equivalent to believing that a tornado could sweep through a junkyard and form a Boeing 747."

Even with all the correct chemicals present, life cannot be created. If we chopped some existing form of life in a blender, life cannot be generated from all those correct chemicals. Even if we have a real living configuration, we cannot recreate life. Dead or frozen bodies cannot be brought back to life after an extended lifeless time.

We need to accept the explanation that God created the universe and all life and direct our beliefs and actions to be consistent with the acknowledgment of the creator and Jesus' revelation that He is the way, the truth, and the life.

Recent rapid advances in the study of genetics and molecular biology have produced additional insight into fundamental questions such as the origin of life on Earth. This paper provides an overview of Genetics from an engineering perspective as well as discussions of these questions.

The striking similarities are present between genetics processes and computer science. There is a digital genetic code. There is digital logic with "and" functions and logic matrices. There is even "error correcting" in digital copying of genetic code.

GENETICS AND ORIGIN OF LIFE

All living organisms have a genetic code generally represented by the sequence of nucleotides in their DNA. Since there are four possible bases used in construction of the code (denoted A, G, C, and T), each letter in the code carries two bits of information. The DNA is normally in the form of a double strand (the famous "double helix") where the second strand is complementary to the first strand. That is, in the second strand a sequence such as "AGCTTT" is replaced by "TCGAAA" which carries the same information. So the terminology "base pair" refers to one letter of genetic code represented by the base and its complement and equivalent to two bits of information in computer parlance.

Humans have a genetic code (the "genome") of about 3.3 billion base pairs (6.6 gigabits or 825 megabytes). Yes, the human genome would easily fit on a typical laptop hard drive. Each human has two copies of the genome in virtually every cell of his or her body because of inheritance of one set each from both father and mother. Actually males have about 2 percent less code in one of their genomes because they only have one "X" chromosome. During growth and normal life processes cells endlessly read, and interpret the genetic codes, copy various snippets of the code, and use the copies as templates in the manufacture of proteins.

Early naturalists thought that genetic traits were inherited in more or less "analog" fashion in which offspring had an average of their parent's characteristics. Gregor Mendel was the first to realize through extensive experiments with breeding of peas that at the lowest level, inheritance is binary, and that there is a minimum unit of inheritance now known as a "gene". Mendel found that some traits are "recessive" and can appear in progeny even if neither parent has the trait. He also found that inheritance of a trait is independent of inheritance of other traits.

Genes are now known to be implemented as sequences of genetic code that direct specific cells to produce a particular protein at a particular time. An essentially infinite number of possible different protein molecules can be produced depending on the particular order of amino acid molecules used in their construction. The code for protein production has been "broken" so that we now know that a three-letter sequence (a codon) is used to specify a particular amino acid (there are 20 amino acids). For instance, the sequence GGC specifies that the amino acid glycine is to be added to a protein molecule. Start and stop codons mark the beginning and end of a protein coding sequence in a manner startlingly like modern data communications schemes. There are 64 possible codons and only 20 possible amino acids so some redundancy and error correction exists. The regulatory code sequences in genes that specify in which parts of the body and/or at which times a protein will be produced are much more complex and less well understood.

For humans, approximately 45,000 genes are contained in 23 separate strands of DNA known as chromosomes (46 if both sets of code are counted). The number of chromosomes is not indicative of complexity. Dogs have 78; Horses have 64. Ferns have 512. The international Human Genome Project (HGP) has completed a preliminary "draft" sequencing of the entire human genome genetic code. Sequences of a small number of other organisms such as the mouse, fruit fly, and e coli are completed or in work. Having the sequence is very different from understanding what it means.

Mendelian Genetics is somewhat like Newtonian Physics. Eventually scientists noticed subtle deviations from the inheritance model predicted by Mendel. Specifically, inheritance of certain traits was not completely independent of other traits. We now know that inheritance of traits will be independent only if they are carried by different chromosomes and that the probability of jointly inheriting traits carried by the same chromosome is proportional to the physical distance between the two genes on the chromosome. Extensive inheritance studies have resulted in maps of the genome showing the approximate location of some trait genes and human genetic disease genes on specific chromosomes. This information can eventually be combined with the detailed sequence data to disclose the genes which (when incorrect) are responsible for genetic diseases. There are an estimated 3000 different human genetic diseases.

GENE LOGIC

The genetic code has been compared to a blueprint specifying the design of an organism. In fact the genetic code specifies not only the design of the organism but provides for the mechanisms needed to "read" the code and manufacture the components of the organism as well as specifying the procedures needed for the life processes of the finished organism. Simple organisms are completely defined genetically. Each tiny nematode worm has exactly 958 cells. Humans, on the other hand, have trillions of cells and less than 100,000 genes so the genetic code is more of a general plan. For example, major blood vessels are genetically specified. Everybody has an aorta. But minor blood vessels grow where needed according to genetically defined rules.

Although all the somatic cells in an organism contain the complete genetic code, in any given cell only a relatively few genes are active. The difference in the genes that are active determines the difference between, say, liver and brain cells. A complex gene logic determines when and where a particular gene will be "turned on". The gene logic can accommodate varying amounts of positional detail. The eye, which has a complex structure in which adjacent cells can be very different, presumably requires many genes to implement a relatively small structure. The femur is much larger but much less complex and requires less genetic information. The gene logic also controls when various activities will take place. Cells divide rapidly in growing organisms but do not divide in adults unless needed to replace dead or discarded cells. (Cancer involves a major breakdown in the gene logic in which cells grow in both an inappropriate position and at an inappropriate time. Cancer is thought to require multiple mutations, some of which can be inherited.)

The gene logic is implemented using signaling proteins. That is, genes can control the production of proteins which are actually the building blocks to produce muscle and other structural components in growing cells but can also control production of other proteins which are logic signals. These logic signals can then be received by other genes and determine whether those genes are activated. Some logic proteins are long range in that they can travel through essentially the entire organism (think insulin). Other shorter range signals appear only near their point of origin, possibly only immediately around the cell in which they are generated. Since genes can both generate and detect signaling proteins, many genes can implement a very complex logic. The positional logic framework which governs where in the body specific types of cells are found is an example of a "boot-strap" problem. The logic framework itself has to be constructed as the organism grows from a one-cell fertilized egg to an adult.

A mutation occurs when the genetic code in a cell is altered such that descendent cells formed by division of the altered cell also have the altered DNA. If the mutation occurs in the chain of cell division between the original fertilized egg and reproductive (sperm or egg) cells (the germ line), then the mutation can be passed to progeny. Many other mutations presumably have no effect because they occur in genes that are never activated in the descendents of the affected cells. (A mutation in a gene that was only active in the brain would have no effect if it occurred in the line of cells that were to form a leg, etc)

Evolution takes place by means of mutations which affect the germ line. Often a mutation results in loss of some essential function and is therefore fatal to progeny and not passed on to living descendants. Sometimes the mutation results in an evolutionary advantage and therefore may eventually become universal in descendants. Sometimes mutation results in characteristics which are different (such as a red eye colour in a species that previously had only brown eyes) but confers no particular advantage or disadvantage and so becomes common but not universal in descendants. Higher organisms also have extensive non-functional sections in their genetic code. Mutations in the non-functional parts of the code would have no observable effect on the organism and therefore would be passed to descendents. Since the rate at which mutations occur should be relatively constant, differences in non-functional code can be used to determine the time since two individuals shared a common ancestor.

The HGP indicates that the human genome contains about 50 percent apparently non-functional code much of which consists of many repetitions of simple sequences such as ..ATATATATATAT...which clearly have little or no information content. Some of the repeat sequences are known to be necessary synchronization patterns such as the sequences at the beginnings and ends of chromosomes. The purpose, if any, of other repeat sequences is unknown.

ORIGIN OF LIFE ON EARTH

So, what does all this have to do with the origin of life?

The genetic code represents an historical record of the development of the organism with an extraordinary amount of detail (825 megabytes is a lot of detail!). An organism which shares significant code sequences with another organism very likely has a common ancestor. By looking at changes in non-functional DNA we can estimate the time since that ancestor lived. By comparing genomes we can construct a "family tree" of life on Earth.

Based on data from the HGP and other sources we can say things like the following:

- All humans are descended from a single individual who lived about 270,000 years ago.
- Humans and New World monkeys share an ancestor which lived about 7 million years ago.
- Humans and mice share a common ancestor which lived about 50 million years ago.
- All life on earth is thought to be descended from an original primordial single cell organism which lived about 3.5 billion years ago.
- The Earth was formed about 4.5 billion years ago but was probably incompatible with life until perhaps 3.8 billion years ago so life apparently appeared relatively quickly.

As more genetic code data is available on various other organisms and as analysis of differences and similarities of codes progresses the entire family tree of life on earth will eventually be developed and more will be known about the characteristics of the primordial organism.

SO, where did that original primordial organism come from?

There seem to be several schools of thought.

THEORY 1 - LIFE APPEARED SPONTANEOUSLY

Some scientists believe that life arose spontaneously from available materials present on the early Earth. In fact, experiments have been conducted in which air, water, carbon dioxide, methane, and common minerals were "cooked" in the presence of energy sources such as heat, sunlight, and simulated lightning to see if life or precursors of life would appear. Indeed, organic "building blocks" such as amino acids did appear.

But it is a very, very long way from amino acids to a life form. The genetic work indicates that the complexity of genetic codes doesn't track that well with the apparent complexity of the organism and that even very simple organisms have quite complex genomes. The simplest known living thing is the microbe mycoplasma genitalium which causes human non-gonococcal urethritus. This microbe has a genetic code of about 570,000 base pairs. Viruses are simpler but aren't really "alive" in the sense that they cannot reproduce or grow without using the mechanisms in a living cell to do so. The bacteria e coli has a genetic code of about 5.7 million base pairs.

But e coli and mycoplasma can't live in the absence of other more complex organisms (e coli lives in animal gut, mycoplasma lives in ... well you get the idea). In fact the primordial organism must have been at the bottom of the food chain, capable of synthesizing its own food from non-living material, and living without assistance from any other living organism. It could have possibly been something on the order of blue-green algae which has 3.6 million base pairs in its genetic code and is thought to be about 3.5 billion years old. Mycoplasma, bacteria, and viruses all must have "devolved" from more complex organisms in response to the availability of more complex forms to act as hosts or links in the food chain.

The original organism had mechanisms (ability to grow, reproduce, and evolve) which led to the evolution of the diverse life forms which now exist on Earth and as indicated above this evolution is documented in the genetic codes of organisms now alive as well as in fossil evidence. But under this scenario the original organism would have had to appear by random happenstance aggregation of materials. This is somewhat like believing that because while digging you found a rock that looked like a brick, if you dug long and hard enough you would eventually find something that looked like the Sistine Chapel complete with Michelangelo's Creation on the ceiling.

Life Evolved from Simpler Organisms

Some scientists feel that life originated spontaneously as a much simpler organism than any now found. A difficulty with this idea is the absence of any current examples of the simpler organism. In general, appearance of more complex organisms has not resulted in disappearance of simpler forms. We still have cockroaches. We still have fruit flies. Indeed, there is evidence of devolution. Eventually, back tracking of many living genetic codes should enable some insight into the probability of this theory.

Its Unknowable

Some scientists take the view that the origin of the primordial organism is "unknowable" meaning that not only do we not know but we are unlikely to ever know and that the subject is therefore more appropriate for philosophy or religion than science. The origin of the primordial organism is therefore the biological equivalent of the "Big Bang Theory" in Astrophysics in which astrophysicists think the entire universe was once the size of a golf ball which then exploded to create the observed universe. They can trace observed cosmic phenomena such as galaxies, red shift, and background radiation back to the golf ball but they admit that it is "unknowable" as to how the golf ball got there.

It Came from Outer Space

Some believe that life originated elsewhere in the universe and was then somehow distributed. This doesn't have to mean biological contamination of the early Earth by space travelers flushing their ballast tanks. It could be that life was distributed via simple frozen or sporelated organisms carried by fragments of a destroyed planet or ejecta blasted into space by meteorite impact. Comets are known to contain water and ice and NASA thinks it has found evidence of fossilized bacteria in meteorites. DNA has been recovered from material 20 million years old.

The possibility that life originated somewhere else in the universe (it is a Very large universe) and then came here seems to many more likely than the idea that life originated on Earth. The space theory is also less egocentric. Keep in mind that all previous "Earth is the centre of the universe" theories have been disproved.

A consequence of the space theory is that life might be widely distributed. Life might appear relatively rapidly on any planet that has appropriate conditions, at least in regions which were in a position to be seeded from the source – a sort of "universe as Petri dish" concept. In other words, if there is life on Earth, then there is likely to be life in any nearby system that has planets with appropriate conditions.

HISTORY OF THE CONCEPT OF LIFE IN SCIENCE

In a letter to Joseph Dalton Hooker of February 1 1871, Charles Darwin made the suggestion that the original spark of life may have begun in a "warm little pond, with all sorts of ammonia and phosphoric salts, lights, heat, electricity, etc. present, so that a protein compound was chemically formed ready to undergo still more complex changes". He went on to explain that "at the present day such matter would be instantly devoured or absorbed, which would not have been the case before living creatures were formed." In other words, the presence of life itself prevents the spontaneous generation of simple organic compounds from occurring on Earth today – a circumstance which makes the search for the origin of life dependent on the sterile conditions of the laboratory. An experimental approach to the question was beyond the scope of laboratory science in Darwin's day, and no real progress was made until 1924 when Aleksandr Ivanovich Oparin demonstrated that it was the presence of atmospheric oxygen and other more sophisticated life-forms that prevented the chain of events that would lead to the evolution of life.

Fig. Charles Darwin

In his The Origin of Life on Earth, Oparin argued that a "primeval soup" of organic molecules could be created in an oxygen-less atmosphere through the action of sunlight. These would combine in ever-more complex fashion until they dissolved into a coacervate droplet. These droplets would "grow" by fusion with other droplets, and "reproduce" through fission into daughter droplets, and so have a primitive metabolism in which those factors which promote "cell integrity" survive, those that don't become extinct. All modern theories of the origin of life take Oparin's ideas as a starting point.

CURRENT MODELS ON ORIGIN OF LIFE

There is no truly "standard" model of the origin of life. But most currently accepted models build in one way or another upon a number of discoveries about the origin of molecular and cellular components for life, which are listed in a rough order of postulated emergence:

Plausible pre-biotic conditions result in the creation of certain basic small molecules (monomers) of life, such as amino acids. This was demonstrated in the Miller-Urey experiment by Stanley L. Miller and Harold C. Urey in 1953.

Phospholipids (of an appropriate length) can spontaneously form lipid bilayers, a basic component of the cell membrane.

The polymerization of nucleotides into random RNA molecules might have resulted in self-replicating ribozymes (RNA world hypothesis).

Selection pressures for catalytic efficiency and diversity result in ribozymes which catalyse peptidyl transfer (hence formation of small proteins), since oligopeptides complex with RNA to form better catalysts. Thus the first ribosome is born, and protein synthesis becomes more prevalent.

Proteins outcompete ribozymes in catalytic ability, and therefore become the dominant biopolymer. Nucleic acids are restricted to predominantly genomic use.

The origin of the basic biomolecules, while not settled, is less controversial than the significance and order of steps 2 and 3. The basic chemicals from which life was thought to have formed are methane (CH_4), ammonia (NH_3), water (H_2O), hydrogen sulfide (H_2S), carbon dioxide (CO_2) or carbon monoxide (CO), and phosphate (PO_4^{3-}). Molecular oxygen (O_2) and ozone (O_3) were either rare or absent.

As 2007, no one has yet synthesized a "protocell" using basic components which would have the necessary properties of life (the so-called "bottom-up-approach"). Without such a proof-of-principle, explanations have tended to be short on specifics. However, some researchers are working in this field, notably Steen Rasmussen at Los Alamos National Laboratory and Jack Szostak at Harvard University. Others have argued that a "top-down approach" is more feasible. One such approach, attempted by Craig Venter and others at The Institute for Genomic Research, involves engineering existing prokaryotic cells with progressively fewer genes, attempting to discern at which point the most minimal requirements for life were reached. The biologist John Desmond Bernal, coined the term Biopoesis for this process, and suggested that there were a number of clearly defined "stages" that could be recognised in explaining the origin of life.

Stage 1: The origin of biological monomers

Stage 2: The origin of biological polymers

Stage 3: The evolution from molecules to cell

Bernal suggested that Darwinian evolution may have commenced early, some time between Stage 1 and 2.

ORIGIN OF ORGANIC MOLECULES

The Miller-Urey experiment attempted to recreate the chemical conditions of the primitive Earth in the laboratory, and synthesized some of the building blocks of life.

MILLER'S EXPERIMENTS

In 1953 a graduate student, Stanley Miller, and his professor, Harold Urey, performed an experiment that proved organic molecules could have spontaneously formed on Early Earth from inorganic precursors. The now-famous "Miller-Urey experiment" used a highly reduced mixture of gases - methane, ammonia and hydrogen – to form basic organic monomers, such as amino acids. Whether the mixture of gases used in the Miller-Urey experiment truly reflects the atmospheric content of Early Earth is a controversial topic. Other less reducing gases produce a lower yield and variety. It was once thought that appreciable amounts of molecular oxygen were present in the prebiotic atmosphere, which would have essentially prevented the formation of organic molecules; however, the current scientific consensus is that such was not the case.

A recent experiment showed that a thick organic haze might have blanketed Early Earth. An organic haze can form over a wide range of methane and carbon dioxide concentrations, believed to be present in the atmosphere of Early Earth. After forming, these organic molecules would have floated down all over the Earth, allowing life to flourish globally.

Simple organic molecules are of course a long way from a fully functional self-replicating life form. But in an environment with no pre-existing life these molecules may have accumulated and provided a rich environment for chemical evolution ("soup theory"). On the other hand, the spontaneous formation of complex polymers from abiotically generated monomers under these conditions is not at all a straightforward process. Besides the necessary basic organic monomers, also compounds that would have prohibited the formation of polymers were formed in high concentration during the experiments.

Other sources of complex molecules have been postulated, including sources of extra-terrestrial stellar or interstellar origin. For example, from spectral analyses, organic molecules are known to be present in comets and meteorites. In 2004, a team detected traces of polycyclic aromatic hydrocarbons (PAH's) in a nebula, the most complex molecule, to that date, found in space. The use of PAH's has also been proposed as a precursor to the RNA world in the PAH world hypothesis.

It can be argued that the most crucial challenge unanswered by this theory is how the relatively simple organic building blocks polymerise and form more complex structures, interacting in consistent ways to form a protocell. For example, in an aqueous environment hydrolysis of oligomers/polymers into their constituent monomers would be favored over the condensation of individual monomers into polymers. Also, the Miller experiment produces many substances that would undergo cross-reactions with the amino acids or terminate the peptide chain.

EIGEN'S HYPOTHESIS

In the early 1970s a major attack on the problem of the origin of life was organised by a team of scientists gathered around Manfred Eigen of the Max Planck Institute. They tried to examine the transient stages between the molecular chaos in a prebiotic soup and the transient stages of a self replicating hypercycle, between the molecular chaos in a prebiotic soup and simple macromolecular self-reproducing systems.

In a hypercycle, the information storing system (possibly RNA) produces an enzyme, which catalyzes the formation of another information system, in sequence until the product of the last aids in the formation of the first information system. Mathematically treated, hypercycles could create quasispecies, which through natural selection entered into a form of Darwinian evolution. A boost to hypercycle theory was the discovery that RNA, in certain circumstances forms itself into ribozymes, a form of RNA enzyme.

WÄCHTERSHÄUSER'S HYPOTHESIS

Another possible answer to this polymerization conundrum was provided in 1980s by Günter Wächtershäuser, in his iron-sulphur world theory. In this theory, he postulated the evolution of (bio)chemical pathways as fundamentals of the evolution of life. Moreover, he presented a consistent system of tracing today's biochemistry back to ancestral reactions that provide alternative pathways to the synthesis of organic building blocks from simple gaseous compounds.

In contrast to the classical Miller experiments, which depend on external sources of energy (such as simulated lightning or UV irradiation), "Wächtershäuser systems" come with a built-in source of energy, sulfides of iron and other minerals (e.g. pyrite). The energy released from redox reactions of these metal sulfides is not only available for the synthesis of organic molecules, but also for the formation of oligomers and polymers. It is therefore hypothesized that such systems may be able to evolve into autocatalytic sets of self-replicating, metabolically active entities that would predate the life forms known today.

The experiment as performed, produced a relatively small yield of dipeptides (0.4% to 12.4%) and a smaller yield of tripeptides (0.003%) and the authors note that: "under these same conditions dipeptides hydrolysed rapidly." Another criticism of the result is that the experiment did not include any organomolecules that would most likely cross-react or chain-terminate.

The latest modification of the iron-sulphur-hypothesis was provided by William Martin and Michael Russell in 2002. According to their scenario, the first cellular life forms may have evolved inside so-called black smokers at seafloor spreading zones in the deep sea. These structures consist of microscale caverns that are coated by thin membraneous metal sulfide walls. Therefore, these structures would solve several critical points of the "pure" Wächtershäuser systems at once:

The micro-caverns provide a means of concentrating newly synthesised molecules, thereby increasing the chance of forming oligomers;

The steep temperature gradients inside a black smoker allow for establishing "optimum zones" of partial reactions in different regions of the black smoker (e.g. monomer synthesis in the hotter, oligomerisation in the colder parts);

the flow of hydrothermal water through the structure provides a constant source of building blocks and energy (freshly precipitated metal sulfides);

the model allows for a succession of different steps of cellular evolution (prebiotic chemistry, monomer and oligomer synthesis, peptide and protein synthesis, RNA world, ribonucleoprotein assembly and DNA world) in a single structure, facilitating exchange between all developmental stages;

synthesis of lipids as a means of "closing" the cells against the environment is not necessary, until basically all cellular functions are developed.

This model locates the "last universal common ancestor" (LUCA) inside a black smoker, rather than assuming the existence of a free-living form of LUCA. The last evolutionary step would be the synthesis of a lipid membrane that finally allows the organisms to leave the microcavern system of the black smokers and start their independent lives. This postulated late acquisition of lipids is consistent with the presence of completely different types of membrane lipids in archaebacteria and eubacteria (plus eukaryotes) with highly similar cellular physiology of all life forms in most other aspects.

Another unsolved issue in chemical evolution is the origin of homochirality, i.e. all monomers having the same "handedness" (amino acids being left handed, and nucleic acid sugars being right handed). Homochirality is essential for the formation of functional ribozymes (and probably proteins too). The origin of homochirality might simply be explained by an initial asymmetry by chance followed by common descent. Work performed in 2003 by scientists at Purdue identified the amino acid serine as being a probable root cause of organic molecules' homochirality. Serine forms particularly strong bonds with amino acids of the same chirality, resulting in a cluster of eight molecules that must be all right-handed or left-handed. This property stands in contrast with other amino acids which are able to form weak bonds with amino acids of opposite chirality. Although the mystery of why left-handed serine became dominant is still unsolved, this result suggests an answer to the question of chiral transmission: how organic molecules of one chirality maintain dominance once asymmetry is established.

"GENES FIRST" MODELS: THE RNA WORLD

The RNA world hypothesis suggests that relatively short RNA molecules could have spontaneously formed that were capable of catalyzing their own continuing replication. It is difficult to gauge the probability of this formation. A number of theories of modes of formation have been put forward. Early cell membranes could have formed spontaneously from proteinoids, protein-like molecules that are produced when amino acid solutions are heated - when present at the correct concentration in aqueous solution, these form microspheres which are observed to behave similarly to membrane-enclosed compartments. Other possibilities include systems of chemical reactions taking place within clay substrates or on the surface of pyrite rocks. Factors supportive of an important role for RNA in early life include its ability to replicate; its ability to act both to store information and catalyse chemical reactions (as a ribozyme); its many important roles as an intermediate in the expression and maintenance of the genetic information (in the form of DNA) in modern organisms; and the ease of chemical synthesis of at least the components of the molecule under conditions approximating the early Earth.

A number of problems with the RNA world hypothesis remain, particularly the instability of RNA when exposed to ultraviolet light, the difficulty of activating and ligating nucleotides and the lack of available phosphate in solution required to constitute the backbone, and the instability of the base cytosine (which is prone to hydrolysis). Recent experiments also suggest that the original estimates of the size of an RNA molecule capable of self-replication were most probably vast underestimates. More-modern forms of the RNA World theory propose that a simpler molecule was capable of self-replication (that other "World" then evolved over time to produce the RNA World). At this time however, the various hypotheses have incomplete evidence supporting them. Many of them can be simulated and tested in the lab, but a lack of undisturbed sedimentary rock from that early in Earth's history leaves few opportunities to test this hypothesis robustly.

"METABOLISM FIRST" MODELS: IRON-SULPHUR WORLD AND OTHERS

Several models reject the idea of the self-replication of a "naked-gene" and postulate the emergence of a primitive metabolism which could provide an environment for the later emergence of RNA replication.

One of the earliest incarnations of this idea was put forward in 1924 with Alexander Oparin's notion of primitive self-replicating vesicles which predated the discovery of the structure of DNA. More recent variants in the 1980s and 1990s include Günter Wächtershäuser's iron-sulphur world theory and models introduced by Christian de Duve based on the chemistry of thioesters. More abstract and theoretical arguments for the plausibility of the emergence of metabolism without the presence of genes include a mathematical model introduced by Freeman Dyson in the early 1980s and Stuart Kauffman's notion of collectively autocatalytic sets, discussed later in that decade.

However, the idea that a closed metabolic cycle, such as the reductive citric acid cycle, could form spontaneously (proposed by Günter Wächtershäuser) remains unsupported. According to Leslie Orgel, a leader in origin-of-life studies for the past several decades, there is reason to believe the assertion will remain so. Orgel summarizes his analysis of the proposal by stating, "There is at present no reason to expect that multistep cycles such as the reductive citric acid cycle will self-organize on the surface of FeS/FeS2 or some other mineral." It is possible that another type of metabolic pathway was used at the beginning of life. For example, instead of the reductive citric acid cycle, the "open" acetyl-CoA pathway (another one of the four recognised ways of carbon dioxide fixation in nature today) would be even more compatible with the idea of self-organisation on a metal sulfide surface. The key enzyme of this pathway, carbon monoxide dehydrogenase/acetyl-CoA synthase harbours mixed nickel-iron-sulphur clusters in its reaction centres and catalyses the formation of acetyl-CoA (which may be regarded as a modern form of acetyl-thiol) in a single step.

BUBBLE THEORY

Waves breaking on the shore create a delicate foam composed of bubbles. Winds sweeping across the ocean have a tendency to drive things to shore, much like driftwood collecting on the beach. It is possible that organic molecules were concentrated on the shorelines in much the same way. Shallow coastal waters also tend to be warmer, further concentrating the molecules through evapouration. While bubbles comprised of mostly water burst quickly, oily bubbles happen to be much more stable, lending more time to the particular bubble to perform these crucial experiments.

The phospholipid is a good example of an oily compound believed to have been prevalent in the prebiotic seas. Because phospholipids contain a hydrophilic head on one end, and a hydrophobic tail on the other, they have the tendency to spontaneously form lipid membranes in water. A lipid monolayer bubble can only contain oil, and is therefore not conducive to harbouring water-soluble organic molecules. On the other hand, a lipid bilayer bubble can contain water, and was a likely precursor to the modern cell membrane. If a protein came along that increased the integrity of its parent bubble, then that bubble had an advantage, and was placed at the top of the natural selection waiting list. Primitive reproduction can be envisioned when the bubbles burst, releasing the results of the experiment into the surrounding medium. Once enough of the 'right stuff' was released into the medium, the development of the first prokaryotes, eukaryotes, and multicellular organisms could be achieved. This theory is expanded upon in the book, "The Cell: Evolution of the First Organism" by Joseph Panno.

Similarly, bubbles formed entirely out of protein-like molecules, called microspheres, will form spontaneously under the right conditions. But they are not a likely precursor to the modern cell membrane, as cell membranes are composed primarily of lipid compounds rather than amino-acid compounds (for types of membrane spheres associated with abiogenesis,see protobionts, micelle, coacervate).

Hybrid Models

A growing realization of the inadequacy of either pure "genes-first" or "metabolism-first" models is leading the trend towards models that incorporate aspects of each.

OTHER MODELS

AUTOCATALYSIS

British ethologist Richard Dawkins wrote about autocatalysis as a potential explanation for the origin of life in his 2004 book. The Ancestor's Tale. Autocatalysts are substances which catalyze the production of themselves, and therefore have the property of being a simple molecular replicator. In his book, Dawkins cites experiments performed by Julius Rebek and his colleagues at the Scripps Research Institute in California in which they combined amino adenosine and pentafluorophenyl ester with the autocatalyst amino adenosine triacid ester (AATE). One system from the experiment contained variants of AATE which catalysed the synthesis of themselves. This experiment demonstrated the possibility that autocatalysts could exhibit competition within a population of entities with heredity, which could be interpreted as a rudimentary form of natural selection.

CLAY THEORY

A hypothesis for the origin of life based on clay was forwarded by Dr A. Graham Cairns-Smith of the University of Glasgow in 1985 and adopted as a plausible illustration by just a handful of other scientists (including Richard Dawkins). Clay theory postulates that complex organic molecules arose gradually on a pre-existing, non-organic replication platform – silicate crystals in solution. Complexity in companion molecules developed as a function of selection pressures on types of clay crystal is then exapted to serve the replication of organic molecules independently of their silicate "launch stage". It is, truly, "life from a rock."

Cairns-Smith is a staunch critic of other models of chemical evolution. However, he admits, that like many models of the origin of life, his own also has its shortcomings

Peggy Rigou of the National Institute of Agronomic Research (INRA), in Jouy-en-Josas, France reports that prions are capable of binding to clay particles and migrate off the particles when the clay becomes negatively charged. While no reference is made in the report to implications for origin-of-life theories, this research may suggest prions as a likely pathway to early reproducing molecules.

"DEEP-HOT BIOSPHERE" MODEL OF GOLD

The discovery of nanobes (filamental structures smaller than bacteria containing DNA) in deep rocks, led to a controversial theory put forward by Thomas Gold in the 1990s that life first developed not on the surface of the Earth, but several kilometers below the surface. It is now known that microbial life is plentiful up to five kilometers below the earth's surface in the form of archaea, which are generally considered to have originated either before or around the same time as eubacteria, most of which live on the surface including the oceans. It is claimed that discovery of microbial life below the surface of another body in our solar system would lend significant credence to this theory. He also noted that a trickle of food from a deep, unreachable, source promotes survival because life arising in a puddle of organic material is likely to consume all of its food and become extinct.

"PRIMITIVE" EXTRATERRESTRIAL LIFE

An alternative to Earthly abiogenesis is the hypothesis that primitive life may have originally formed extraterrestrially, either in space or on a nearby planet (Mars). (Note that exogenesis is related to, but not the same as, the notion of panspermia).

Organic compounds are relatively common in space, especially in the outer solar system where volatiles are not evapourated by solar heating. Comets are encrusted by outer layers of dark material, thought to be a tar-like substance composed of complex organic material formed from simple carbon compounds after reactions initiated mostly by irradiation by ultraviolet light. It is supposed that a rain of material from comets could have brought significant quantities of such complex organic molecules to Earth.

An alternative but related hypothesis, proposed to explain the presence of life on Earth so soon after the planet had cooled down, with apparently very little time for prebiotic evolution, is that life formed first on early Mars. Due to its smaller size Mars cooled before Earth(a difference of hundreds of millions of years), allowing prebiotic processes there while Earth was still too hot. Life was then transported to the cooled Earth when crustal material was blasted off Mars by asteroid and comet impacts. Mars continued to cool faster and eventually became hostile to the continued evolution or even existence of life (it lost its atmosphere due to low volcanism), Earth is following the same fate as Mars, but at a slower rate.

Neither hypothesis actually answers the question of how life first originated, but merely shifts it to another planet or a comet. However, the advantage of an extraterrestrial origin of primitive life is that life is not required to have evolved on each planet it occurs on, but rather in a single location, and then spread about the galaxy to other star systems via cometary and/or meteorite impact. Evidence to support the plausibility of the concept is scant, but it finds support in recent study of Martian meteorites found in Antarctica and in studies of extremophile microbes. Additional support comes from a recent discovery of a bacterial ecosytem whose energy source is radioactivity.

Chapter 4

Anatomy

Anatomy (from the Greek word anatome,"dissection"), is a branch of natural science dealing with the structural organization of living things. It is an old science, having its beginnings in prehistoric times. For centuries anatomical knowledge consisted largely of observations of dissected plants and animals. The proper understanding of structure, however, implies a knowledge of function in the living organism. Anatomy is therefore almost inseparable from physiology, which is sometimes called functional anatomy. As one of the basic life sciences, anatomy is closely related to medicine and to other branches of biology.

It is convenient to subdivide the study of anatomy in several different ways. One classification is based on the type of organisms studied, the major subdivisions being plant anatomy and animal anatomy. Animal anatomy is further subdivided into human anatomy (which we cover in detail) and comparative anatomy, which seeks out similarities and differences among animal types. Anatomy can also be subdivided into biological processes—for example, developmental anatomy, the study of embryos, and pathological anatomy, the study of diseased organs. Other subdivisions, such as surgical anatomy and anatomical art, are based on the relationship of anatomy to other branches of activity under the general heading of applied anatomy. Still another way to subdivide anatomy is by the techniques employed—for example, microanatomy, which concerns itself with observations made with the help of the microscope (see the section below on the history of anatomy).

HUMAN ANATOMY

The human body operates through systems that are summarized below. This web site contains separate articles on each of the systems and organs mentioned to which the reader is referred for more comprehensive discussions.

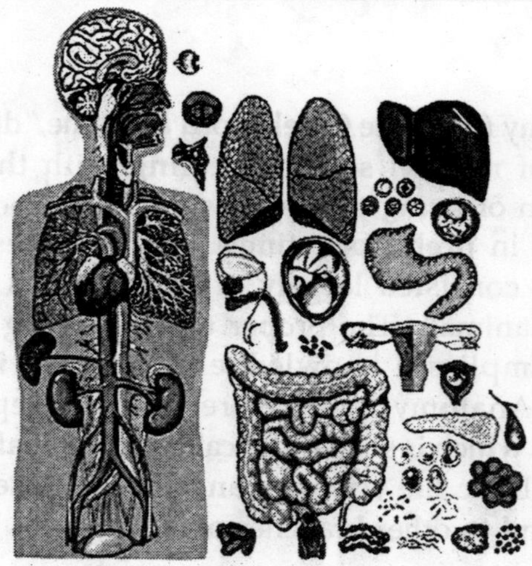

Fig. Human Body

Musculoskeletal System

The human skeleton consists of more than 200 bones bound together by tough and relatively inelastic connective tissues called ligaments.

The different parts of the body vary greatly in their degree of movement. Thus, the arm at the shoulder is freely movable, whereas the knee joint is definitely limited to a hingelike action. The movements of individual vertebrae are extremely limited; the bones composing the skull are immovable. Movements of the bones of the skeleton are effected by contractions of the skeletal muscles, to which the bones are attached by tendons. These muscular contractions are controlled by the nervous system.

Nervous System

The nervous system has two divisions: the somatic, which allows voluntary control over skeletal muscle, and the autonomic, which is involuntary and controls cardiac and smooth muscle and glands. The autonomic nervous system has two divisions: the sympathetic and the parasympathetic. Many, but not all, of the muscles and glands that distribute nerve impulses to the larger interior organs possess a double nerve supply; in such cases the two divisions may exert opposing effects. Thus, the sympathetic system increases heartbeat, and the parasympathetic system decreases heartbeat. The two nervous systems are not always antagonistic, however. For example, both nerve supplies to the salivary glands excite the cells of secretion. Furthermore, a single division of the autonomic nervous system may both excite and inhibit a single effector, as in the sympathetic supply to the blood vessels of skeletal muscle. Finally, the sweat glands, the muscles that cause involuntary erection or bristling of the hair, the smooth muscle of the spleen, and the blood vessels of the skin and skeletal muscle are actuated only by the sympathetic division.

Voluntary movement of head, limbs, and body is caused by nerve impulses arising in the motor area of the cortex of the brain and carried by cranial nerves or by nerves that emerge from the spinal cord to connect with skeletal muscles. The reaction involves both excitation of nerve cells stimulating the muscles involved and inhibition of the cells that stimulate opposing muscles. A nerve impulse is an electrical change within a nerve cell or fiber; measured in millivolts, it lasts a few milliseconds and can be recorded by electrodes.

Movement may occur also in direct response to an outside stimulus; thus, a tap on the knee causes a jerk, and a light shone into the eye makes the pupil contract. These involuntary responses are called reflexes. Various nerve terminals called receptors constantly send impulses into the central nervous system. These are of three classes: exteroceptors, which are sensitive to pain, temperature, touch, and pressure;

interoceptors, which react to changes in the internal environment; and proprioceptors, which respond to variations in movement, position, and tension. These impulses terminate in special areas of the brain, as do those of special receptors concerned with sight, hearing, smell, and taste.

Muscular contractions do not always cause actual movement. A small fraction of the total number of fibers in most muscles are usually contracting. This serves to maintain the posture of a limb and enables the limb to resist passive elongation or stretch. This slight continuous contraction is called muscle tone.

Circulatory System

In passing through the system, blood pumped by the heart follows a winding course through the right chambers of the heart, into the lungs, where it picks up oxygen, and back into the left chambers of the heart. From these it is pumped into the main artery, the aorta, which branches into increasingly smaller arteries until it passes through the smallest, known as arterioles. Beyond the arterioles, the blood passes through a vast amount of tiny, thin-walled structures called capillaries. Here, the blood gives up its oxygen and its nutrients to the tissues and absorbs from them carbon dioxide and other waste products of metabolism. The blood completes its circuit by passing through small veins that join to form increasingly larger vessels until it reaches the largest veins, the inferior and superior venae cavae, which return it to the right side of the heart. Blood is propelled mainly by contractions of the heart; contractions of skeletal muscle also contribute to circulation. Valves in the heart and in the veins ensure its flow in one direction.

Immune System

The body defends itself against foreign proteins and infectious microorganisms by means of a complex dual system that depends on recognizing a portion of the surface pattern of the invader. The two parts of the system are termed cellular immunity, in which lymphocytes are the effective agent, and humoral immunity, based on the action of antibody molecules. When particular lymphocytes recognize a foreign molecular pattern (termed an antigen), they release antibodies in great numbers; other lymphocytes store the memory of the pattern for future release of antibodies should the molecule reappear. Antibodies attach themselves to the antigen and in that way mark them for destruction by other substances in the body's defence arsenal. These are primarily complement, a complex of enzymes that make holes in foreign cells, and phagocytes, cells that engulf and digest foreign matter. They are drawn to the area by chemical substances released by activated lymphocytes.

Lymphocytes, which resemble blood plasma in composition, are manufactured in the bone marrow and multiply in the thymus and spleen. They circulate in the bloodstream, penetrating the walls of the blood capillaries to reach the cells of the tissues. From there they migrate to an independent network of capillaries that is comparable to and almost as extensive as that of the blood's circulatory system. The capillaries join to form larger and larger vessels that eventually link up with the bloodstream through the jugular and subclavian veins; valves in the lymphatic vessels ensure flow in one direction. Nodes at various points in the lymphatic network act as stations for the collection and manufacture of lymphocytes; they may become enlarged during an infectious disease. In anatomy, the network of lymphatic vessels and the lymph nodes are together called the lymphatic system; its function as the vehicle of the immune system was not recognized until the 1960s.

Respiratory System

Respiration is carried on by the expansion and contraction of the lungs; the process and the rate at which it proceeds are controlled by a nervous centre in the brain. In the lungs, oxygen enters tiny capillaries, where it combines with hemoglobin in the red blood cells and is carried to the tissues. Simultaneously, carbon dioxide, which entered the blood in its passages through the tissues, passes through capillaries into the air contained within the lungs. Inhaling draws into the lungs air that is higher in oxygen and lower in carbon dioxide; exhaling forces from the lungs air that is high in carbon dioxide and low in oxygen. Changes in the size and gross capacity of the chest are controlled by contractions of the diaphragm and of the muscles between the ribs.

Digestive and Excretory Systems

The energy required for maintenance and proper functioning of the human body is supplied by food. After it is broken into fragments by chewing (see Teeth) and mixed with saliva, digestion begins. The food passes down the gullet into the stomach, where the process is continued by the gastric and intestinal juices. Thereafter, the mixture of food and secretions, called chyme, is pushed down the alimentary canal by peristalsis, rhythmic contractions of the smooth muscle of the gastrointestinal system. The contractions are initiated by the parasympathetic nervous system; such muscular activity can be inhibited by the sympathetic nervous system. Absorption of nutrients from chyme occurs mainly in the small intestine; unabsorbed food and secretions and waste substances from the liver pass to the large intestines and are expelled as feces. Water and water-soluble substances travel via the bloodstream from the intestines to the kidneys, which absorb all the constituents of the blood plasma except its proteins. The kidneys return most of the water and salts to the body, while excreting other salts and waste products, along with excess water, as urine.

The Endocrine System

In addition to the integrative action of the nervous system, control of various body functions is exerted by the endocrine glands. An important part of this system, the pituitary, lies at the base of the brain. This master gland secretes a variety of hormones, including the following: (1) a hormone that stimulates the thyroid gland and controls its secretion of thyroxine, which dictates the rate at which all cells utilize oxygen; (2) a hormone that controls the secretion in the adrenal gland of hormones that influence the metabolism of carbohydrates, sodium, and potassium and control the rate at which substances are exchanged between blood and tissue fluid; (3) substances that control the secretion in the ovaries of estrogen and progesterone and the creation in the testicles of testosterone; (4) the somatotropic, or growth, hormone, which controls the rate of development of the skeleton and large interior organs through its effect on the metabolism of proteins and carbohydrates; and (5) an insulin inhibitor—a lack of insulin causes diabetes mellitus. The posterior lobe of the pituitary secretes vasopressin, which acts on the kidney to control the volume of urine; a lack of vasopressin causes diabetes insipidus, which results in the passing of large volumes of urine. The posterior lobe also elaborates oxytocin, which causes contraction of smooth muscle in the intestines and small arteries and is used to bring about contractions of the uterus in childbirth. Other glands in the endocrine system are the pancreas, which secretes insulin, and the parathyroid, which secretes a hormone that regulates the quantity of calcium and phosphorus in the blood.

The Reproductive System

Reproduction is accomplished by the union of male sperm and the female ovum. In coitus, the male organ ejaculates more than 250 million sperm into the vagina, from which some make their way to the uterus. Ovulation, the release of an egg into the uterus, occurs approximately every 28 days; during the same period the uterus is prepared for the implantation of a fertilized ovum by the action of estrogens. If a male cell fails to unite with a female cell, other hormones cause the uterine wall to slough off during menstruation. From puberty to menopause, the process of ovulation, and preparation, and menstruation is repeated monthly except for periods of pregnancy. The duration of pregnancy is about 280 days. After childbirth, prolactin, a hormone secreted by the pituitary, activates the production of milk.

Skin

The skin is an organ of double-layered tissue stretched over the surface of the body and protecting it from drying or losing fluid, from harmful external substances, and from extremes of temperature. The inner layer, called the dermis, contains sweat glands, blood vessels, nerve endings (sense receptors), and the bases of hair and nails. The outer layer, the epidermis, is only a few cells thick; it contains pigments, pores, and ducts, and its surface is made of dead cells that it sheds from the body. (Hair and nails are adaptations arising from the dead cells.) The sweat glands excrete waste and cool the body through evapouration of fluid droplets; the blood vessels of the dermis supplement temperature regulation by contracting to preserve body heat and expanding to dissipate it. Separate kinds of receptors convey pressure, temperature, and pain. Fat cells in the dermis insulate the body, and oil glands lubricate the epidermis.

HISTORY OF ANATOMY

The oldest known systematic study of anatomy is contained in an Egyptian papyrus dating from about 1600 BC.

The treatise reveals knowledge of the larger viscera but little concept of their functions. About the same degree of knowledge is reflected in the writings of the Greek physician Hippocrates in the 5th century BC. In the 4th century BC Aristotle greatly increased anatomical knowledge of animals. The first real progress in the science of human anatomy was made in the following century by the Greek physicians Herophilus and Erasistratus, who dissected human cadavers and were the first to distinguish many functions, including those of the nervous and muscular systems. Little further progress was made by the ancient Romans or by the Arabs. The Renaissance first influenced the science of anatomy in the latter half of the 16th century. Modern anatomy began with the publication in 1543 of the work of the Belgian anatomist Andreas Vesalius. Before the publication of this classical work anatomists had been so bound by tradition that the writings of authorities of more than 1000 years earlier, such as the Greek physician Galen, who had been restricted to the dissection of animals, were accepted in lieu of actual observation. Vesalius and other Renaissance anatomists, however, based their descriptions on their own observations of human corpses, thus setting the pattern for subsequent study in anatomy.

MORPHOLOGY

For many years anatomists (even those of the modern era) were concerned mainly with the accumulation of a vast amount of information known as descriptive morphology. Descriptive morphology has been supplemented, and to a certain extent supplanted, by the development of experimental morphology, which attempts to identify the hereditary and environmental determinants in morphology and their relationships by controlled-environment and grafting experiments on embryos. Ideally, anatomical investigation consists of a combination of descriptive and experimental approaches. Present-day anatomy involves scrutiny of the structure of organisms at many levels of observation. For example, the anatomist studies the cells and tissues of organisms with the unaided eye, with simple and compound lenses, with various kinds of microscopes, and by chemical methods of analysis.

ORGANISM

In biology and ecology, an organism is a living complex adaptive system of organs that influence each other in such a way that they function in some way as a stable whole.

The origin of life on Earth and the relationships between its major lineages are controversial. Two main grades may be distinguished, the prokaryotes and eukaryotes. The prokaryotes are generally considered to represent two separate domains, called the Bacteria and Archaea, which are not closer to one another than to the eukaryotes. The gap between prokaryotes and eukaryotes is widely considered a major missing link in evolutionary history. Two eukaryotic organelles, namely mitochondria and chloroplasts, are generally considered to be derived from endosymbiotic bacteria. Fungi, animals and plants are examples of species that are eukaryote. Organisms all go through a life cycle.

The phrase complex organism describes any organism with more than one cell.

The word "organism" may broadly be defined as an assembly of molecules that influence each other in such a way that they function as a more or less stable whole and have properties of life. However, many sources, lexical and scientific, add conditions that are problematic to defining the word.

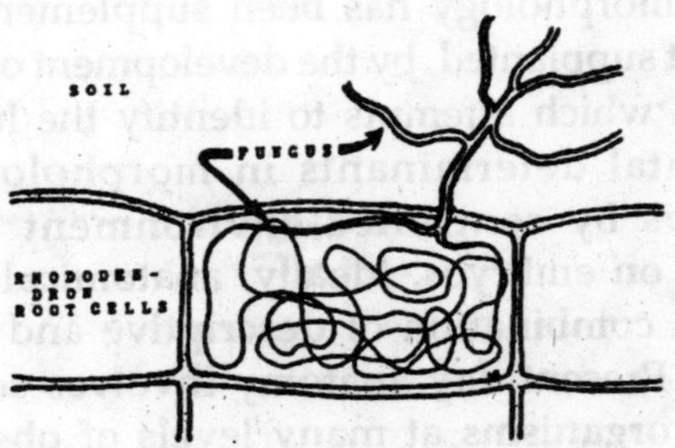

Fig. An ericoid mycorrhizal fungus

The word "organism" usually describes an independent collections of systems (for example circulatory, digestive, or reproductive) themselves collections of organs; these are, in turn, collections of tissues, which are themselves made of cells. The concept of an organism can be challenged on grounds that organisms themselves are never truly independent of an ecosystem; groups or populations of organisms function in an ecosystem in a manner not unlike the function of multicellular tissues in an organism; when organisms enter into strict symbiosis, they are not independent in any sense that could not also be conferred upon an organ or a tissue. Symbiotic plant and algae relationships do consist of radically different DNA structures between contrasting groups of tissues, sufficient to recognize their reproductive independence. However, in a similar way, an organ within an "organism" (say, a stomach) can have an independent and complex interdependent relationship to separate whole organisms, or groups of organisms (a population of viruses, or bacteria), without which the organ's stable function would transform or cease. Other organs within that system (say, the ribcage) might be affected only indirectly by such an arrangement, much the same way species' affect one another indirectly in an ecosystem. Thus, the boundaries of the organism are nearly always disputable, and all living matter exists within larger heterarchical systems of life, made of wide varieties of transient living and dead tissues, and functioning in complex and dynamic relationships to one another.

Viruses

Viruses are not typically considered to be organisms because they are not capable of "independent" reproduction or metabolism. This controversy is problematic, though, since some parasites and endosymbionts are also incapable of independent life. Although viruses have enzymes and molecules characteristic of living organisms, they are incapable of reproducing outside a host cell and most of their metabolic processes require a host and its 'genetic machinery.'

Superorganism

A superorganism is an organism consisting of many organisms. This is usually meant to be a social unit of eusocial animals, where division of labour is highly specialised and where individuals are not able to survive by themselves for extended periods of time. Ants are the most well known example of such a superorganism.

Thermoregulation, a feature usually exhibited by individual organisms, does not occur in individuals or small groups of honeybees of the species Apis mellifera. When these bees pack together in clusters of between 5000 and 40000, the colony can thermoregulate. James Lovelock, with his "Gaia Theory" has paralleled the work of Vladimir Vernadsky, who suggested the whole of the biosphere in some respects can be considered as a superorganism.

Fig. James Lovelock

Classification of Living Things

Scientists have found and described approximately 1.75 million species on Earth. Plus, new species are being discovered every day. From tiny bacteria to yeasts to starfish to blue whales, life's diversity is truly impressive! With such a diversity of life on Earth, how does one go about making sense of it all?

One way to make sense of it is by classification. Scientists put similar species into groups so that those millions of species do not seem so overwhelming. People rely on their knowledge of classification to understand what different species are like. You may have done this without even thinking about it! For instance, let's say that a friend of yours tells you that he saw an egret last weekend. You have never heard of an egret before, but if he tells you that an egret is a type of bird, you should have some idea of what it is like.

Living things are divided into three groups based on their genetic similarity. The three groups are:

Archaea

Archaea are microbes. Most live in extreme environments. These are called extremophyles. Other Archaea species are not extremophiles and live in ordinary temperatures and salinities. Some even live in your guts!

Some extremophile species love the heat! They like to live in boiling water, like the geysers of Yellowstone Park, and inside volcanoes. They like the heat so much that it has earned the nickname "thermophile", which means "loving heat", and it would probably freeze to death at ordinary room temperature. Other extremophile Archaea love to live in very salty, called hypersaline, environments. They are able to survive in these extreme places where other organisms cannot. These salt-loving Archaea are called halophyles.

Archaea was originally thought to be just like bacteria, but archaea is a much different and simpler form of life. It may also be the oldest form of life on Earth!

Archaea requires neither sunlight for photosynthesis as do plants, nor oxygen. Archaea absorbs CO_2, N_2, or H_2S and gives off methane gas as a waste product the same way humans breathe in oxygen and breathe out carbon dioxide. Planets which contain an environment wherein archaea might survive include Venus, the past environment of Mars, Jupiter, Saturn and Jupiter's moon Io.

Eubacteria

More advanced forms of bacteria. Eubacteria, also know as "true bacteria", are microscopic prokaryotic cells. Cyanobacteria, also called blue-green algae, are Eubacteria that have been living on our planet for over 3 billion years. Blue-green algae grow in the shallow parts of the ocean. Today it is only common in certain regions, but a few billion years ago, there was tons of it! Through photosynthesis, which produces oxygen, billions of tiny bacteria were able to add oxygen to Earth's atmosphere. This allowed animals that breath oxygen to survive.

Some Eubacteria can cause health problems like strep throat and food poisoning. Bacteria such as E.coli and Salmonella are sometimes found in undercooked meat and eggs and can make people sick. Other bacteria are good to eat, such as those in yogurt.

People have found that some types of Eubacteria can be very useful. Some are used at wastewater treatment plants to help clean the water. Others are also used to make grapes into wine and milk into cheese.

Eukaryota

All life forms with eukaryotic cells including plants and animals.

What do trees, monkeys, plankton and mushrooms have in common? They are all members of the Eukaryota domain! You are a member of the Eukaryota domain too! Plants, animals, protists, and fungi, are all members of the domain.

These groups might seem pretty different, and they are, but they all have a few key aspects in common including their unique cells that are called eukaryotic. All members of the domain Euraryota have eukaryotic cells. And it is the only domain whose members have this cell type. Eukaryotic cells contain a special part called a nucleus that contains genetic material within chromosomes.

Three Domains of Life

The distance between groups indicates how closely related they are. Groups that are close together, like plants and animals, are much more closely related than groups that are far apart, like plants and bacteria! Do you see how the two types of bacteria, Archaea and Eubacteria, are about as similar to one another as they are to animals? Recent studies have found that bacteria are far more diverse than anyone had suspected. The Eukaryota domain is divided into several groups called kingdoms.

Kingdom Protista: Members of the Kingdom Protista are the simplest of the eukaryotes. Protists are an unusual group of organisms that were put together because they don't really seem to belong to any other group. Some protists perform photosynthesis like plants while others move around and act like animals, but protists are neither plants nor animals. They're not fungi either - even though some might like to "think" they are!

In some ways, the Kingdom Protista is home for the "leftover" organisms that couldn't be classified elsewhere. You might not think a tiny one-celled amoeba has much in common with a giant sea kelp, but they're both members of this kingdom.

Kingdom Fungi: Including mushrooms and other fungus. Though the appearance of many fungi may resemble plants, they are probably more closely related to animals. Fungi are not capable of performing photosynthesis, so must get their nourishment from other sources. Many fungi absorb nutrients directly from t he soil. Many others feed on dead and decaying organisms and therefore have an important role in the recycling of nutrients in natural systems. Still others feed on living organisms. Athlete's foot is a common fungus which feeds on a living host - you!

When you think of fungi, you probably think of the mushrooms we can buy at the supermarket or hunt for in the woods. However, those "mushrooms" are really just special structures called "fruiting bodies" produced by the fungus for reproduction. The rest of the fungus (and the biggest part) lives below the ground. Fungi come in a wide variety of sizes and forms, and many have great economic importance. Tiny, one-celled yeasts are important for baking breads and fermenting wines, beers and vinegars. Many medicines are produced with the help of fungi, most notably, the antibiotic, Penicillin. If you leave your bread on the counter too long, you'll be able to observe a relative of the Penicillium mold for yourself!

Kingdom Plantae: Including trees, grass and flowers. Kingdom Plantae contains almost 300,000 different species of plants. It is not the largest kingdom, but it is a very important one!

In the process known as "photosynthesis", plants use the energy of the Sun to convert water and carbon dioxide into food (sugars) and oxygen. Photosynthesis by plants provides almost all the oxygen in Earth's atmosphere. Because plants can make their own food, they are the first step to many food chains in the world.

The first plants lived on land about 450 million years ago. Since then, plants have taken on many forms and are found in most places on Earth. Plants can live in dry places or wet places, low places or high places, hot places or cold places. Humans can't live in a world without plants, so it is very important to protect places that have plants!

Kingdom Animalia: From snails to birds to mammals like you! With over 2 million species, Kingdom Animalia is the largest of the kingdoms in terms of its species diversity. But when you think of an "animal", what image comes to mind?

While mammals, birds, reptiles, amphibians, and fish are the most familiar to us, over half of all the animal species belong to a group of animals known as arthropods. Arthropods include animals such as centipedes, crabs, insects, and spiders. And with over a million species of arthropods, this means that the majority of animals species come from a group of critters.

So, what exactly is an "animal"? With so much diversity among different animal species, it's difficult to imagine what they all might have in common. First, animals are "multicellular" (composed of many cells). In most animals, these cells are organized into tissues that make up different organs and organ systems. Second, all animals are heterotrophs (= "other feeder"), meaning that they must get their food by eating other organisms, such as plants, fungi, and other animals. (Plants are referred to as autotrophs or "self feeders") because they produce their own food by the process of photosynthesis.) In addition, all animals require oxygen for their metabolism, can sense and respond to their environment, and have the capacity to reproduce sexually (though many reproduce asexually as well). During their development from a fertilized egg to adult, all animals pass through a series of embryonic stages as part of their normal life cycle. Though frogs and humans don't look very much alike, we share many features in common during our embryonic phase.

Within each kingdom, species are further classified into groups based on similarities. For example, the full classification of a human is:

- Domain Eukarya
  - Kingdom Animalia
    - Phylum Chordata
      - Subphylum Vertebrata
        - Class Mammalia
          - Order Primates
            - Family Hominidae
              - Genus Homo
                - Species sapiens

Chapter 5

An Introduction to Cells

Cells are the unit of structure, function, and reproduction in living things. Here is a simpler definition: "Anything that's alive is made up of cells." Every organism that you can think of, whether it be an octopus or a bird, a snail or an ant, is made of cells. Humans are made of literally billions of cells. You can't see them though, because cells are so small that they can only be seen under a microscope. Most organisms, like us, are made up of many different types of cells. Some carry oxygen in our blood, some produce sweat, and some allow us to feel whether it's hot or cold.

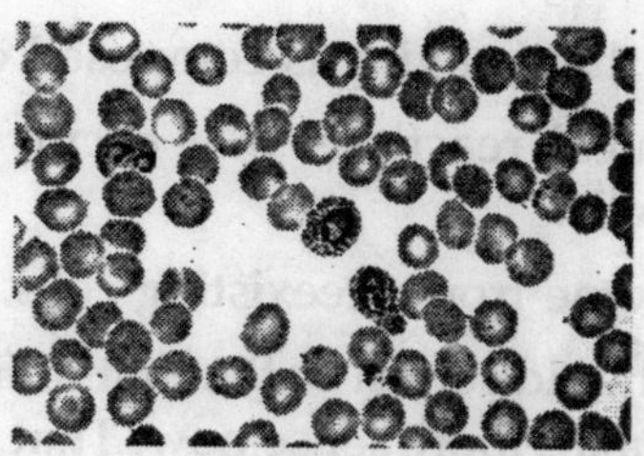

Fig. Red Blood Cells

Although most organisms (an organism is just a living thing) are made of many different types of cells, all of which work together, some organisms are just one, single cell. Biologists call this type of organism unicellular.

Think about it. A unicycle has one wheel, a unicorn has one horn, so an organism that's unicellular has just one cell. Unicellular organisms are very special in that they perform all of the functions necessary to live and yet are so tiny.

Webcytology focuses on how these extraordinary organisms live and function.

The study of cells is known as cytology. Cytology began in 1665, when Robert Hooke, an English scientist, first glimpsed into the microscopic world of cells by examining dead cork cells under a primitive microscope which he constructed. However, all Hooke was able to observe were the thick walls that surrounded each cell. The thought that cells might be the basis for life was not to come for nearly two centuries.

During the next 170 years, other scientists used microscopes to further advance their knowledge of cells. The most important discovery during that period came in 1838, when a German botanist named Matthias Schleiden suggested that all plant tissues are made of cells. Just one year later, zoologist Theodore Schwann made a similar proposal for animals. In 1858, Rudolf Virchow suggested that all cells come from preexisting cells. The ideas of these three scientists led to the creation of what is now called the cell theory. The three main aspects of the theory are:

1. Anything that is living is composed of cells.
2. The chemical reactions that occur in an organism occur in cells.
3. All cells come from preexisting cells.

Perhaps we should return to the first definition of what a cell is to see if it makes more sense now. That definition called a cell the "unit of structure, function, and reproduction in living things." The cell theory says that all living things are composed of cells, so it makes sense that cells are called the unit of structure for an organism. An organism requires chemical reactions to function, and since the second part of the cell theory says that these reactions occur in cells, it can"t be said that cells are the unit of function for the organism. The final component of the theory states that cells reproduce to form new cells. Since all living things reproduce and are also made of cells, the cells themselves must reproduce to form the new organism. That's why the definition also calls cells the unit of reproduction in living things.

Biologists have debated for many decades over how to define what is alive and what isn't; in fact, they are still debating today. How would you define life? Would you say that anything that can grow and reproduce is alive? With that definition, fire would be alive! The "babies" are the sparks that can fly off, which can grow into a completely new fire! To prevent this sort of confusion, biologists have created a list of characteristics that distinguish animate (living) things from inanimate (non-living) things. The most important characteristics are listed below.

1. Living things are in general more organized in their structure than are inanimate objects.
2. Living things reproduce.
3. Living things grow.
4. Living things respond to changes in their surroundings.
5. Living things tend to keep their internal environment unchanged, even when the external environment changes (For example, people have a constant internal temperature of 98.6 degrees Fahrenheit, whether they're in the Sahara Desert or in Antarctica.).

Now return to the example of fire to see why scientists do not consider it to be alive. As mentioned, fire does reproduce and grow, so it satisfies #2 and #3 of above on the list. However, fire does not respond to changes in the environment; if you walk up to a fire, it behaves just as it does if you weren't there, whereas living things (for example, a dog) would exhibit a change in behaviour. Also, fire is not more organized than other inanimate objects. In fact, fire is just the release of energy when something burns, and this release of energy is usually quite random. Finally, while a fire will be hot whether it's in the desert or Antarctica, the fire doesn't really have a well defined internal environment that #6 says it has to have, because fire is just a blob of energy without any real borders.

Fig. A. I. Oparin

You may be wondering, since the cell theory says that all cells come from preexisting cells, where did the first cell come from? The first scientist to suggest a theory which explained the formation of life was the Russian scientist A. I. Oparin.

While certain elements of his theory are still not agreed upon by biologists, most scientists now accept Oparin's ideas regarding the environment of Earth when life first formed, about 3.5 billion years ago. They believe that, at that time, there was very little free oxygen available in the atmosphere (that means that oxygen may have been part of other molecules, like water, but not freely floating in the atmosphere as it is today), and that nitrogen, carbon, oxygen, and hydrogen, were all present in one form or another.

Scientists also generally agree that the Earth was a very violent place at the time when life was first created. Volcanoes, lightning, and ultraviolet rays from the Sun all provided energy that was to forge the first organic molecules. However, this hypothesis was not tested until the 1950s by Stanley Miller, thirty years after Oparin had published his work.

Miller created an apparatus which simulated conditions that scientists believed existed on Earth when life formed. Remarkably, within a mere 24 hours after beginning the experiment, large amounts of amino acid had formed. Subsequent experiments with slightly different initial conditions resulted in the formation of all sorts of other organic molecules. These experiments suggested that the molecules necessary for life could have formed in the primitive Earth. Over the course of time, more and more of them would be produced, until finally they began to chemically combine to form larger molecules. It is thought that these large, organic molecules eventually combined to create the first organisms.

CELLS: ORGINS

The cell is the structural and functional unit of all living organisms, and is sometimes called the "building block of life. Some organisms, such as bacteria, are unicellular, consisting of a single cell. Other organisms, such as humans, are multicellular. (Humans have an estimated 100 trillion or 1014 cells; a typical cell size is 10 μm; a typical cell mass is 1 nanogram.) The largest known cell is an ostrich egg.

The cell theory, first developed in 1839 by Schleiden and Schwann, states that all organisms are composed of one or more cells. All cells come from preexisting cells. Vital functions of an organism occur within cells, and all cells contain the hereditary information necessary for regulating cell functions and for transmitting information to the next generation of cells.

The word cell comes from the Latin cellula, a small room. The name was chosen by Robert Hooke when he compared the cork cells he saw to the small rooms monks lived in.

The origin of cells has to do with the origin of life, and was one of the most important steps in evolution of life as we know it. The birth of the cell marked the passage from prebiotic chemistry to biological life.

The question of how the Earth and everything on it was formed is still being debated. The realm of science is testable, repeatable observations. What do scientists hypothesize about the origins of cells and what evidence do they have to support their ideas?

What follows is brief summary of some of the main ideas held by scientists concerning the creation of the Earth.

The Formation of the Earth

It is believed that the universe began 15 to 20 billion years ago in a "big bang" which created all of the atoms of the universe. Since this is the realm of the theoretical physicists more than the biologist, it will not be discussed here.

Over billions of years, swirling balls of gases collected together to form the stars and planets. A nuclear fusion reaction formed our sun and, around it, nine planets began to condense. About four and a half billion years ago the Earth was a violent mass of molten lava. Geochemists believe the atmosphere was dense with gases such as nitrogen, carbon dioxide, water vapour and possibly methane, ammonia, hydrogen sulfide and hydrogen. As the surface began to cool, the crust formed. Volcanic activity spewed material and water vapour into the air shrouding the Earth in darkness. As the land cooled the water vapour condensed and it began to rain - a rain lasting millions of years. Water covered the surface of the earth - the primordial sea. The waters receded and, about three and a half billion years ago the conditions were right for the beginning of life.

The Formation of Life

Alexander Oparin proposed one of the first hypotheses describing how life could have formed from inorganic molecules. Free oxygen, released into the atmosphere by photosynthetic organisms today, would have been absent in the early atmosphere. Oparin hypothesized that energy from lightning, heat or ultra violet radiation, chemically joined molecules of the atmospheric gases to create simple, carbon-based organic molecules. Rain carried the organic molecules into the seas which became a thick "primordial soup". Within the soup, tiny bubbles of lipid material may have surrounded collections of organic molecules. Some of the droplets eventually may have been able to incorporate new matter from the surrounding seas (grow), and chemically release the energy stored in molecules (metabolize). When a persistent "bubble" collected enough of the right material to grow large enough to divide and pass on to the daughter cells the information they needed to do the same thing (heredity), life had begun.

Ancient rocks from Australia, about 3.5 billion years old contain fossils of the earliest record of living cells. The cells are very simple in structure. They were prokaryotes meaning they had no nucleus and no membrane bound organelles. They were similar to modern day bacteria. Probably the best look at what those first living organisms were like presented itself about 20 years ago with the discovery of an entirely new form of life in the deep sea thermal vents. At depths where no sunlight could reach to allow photosynthesis, bacteria were discovered that could use the chemical energy in superheated plumes of hydrogen sulfide to metabolize and grow. Other organisms such as huge tube worms and blind crabs lived off of the bacteria the way most organisms depend on plants. Methane producing bacteria that are found in hot springs and must live without oxygen (anaerobes) also show similarities to these earliest forms of life.

The Joides Resolution is a research ship whose mission is to drill into and study the sea floor to learn more about the history of the Earth. Some students at Mira Mesa High School recently had the opportunity to tour this ship.

Another simple prokaryotic organism is cyanobacteria. Though lacking membrane bound organelles, these cells have the ability to carry on photosynthesis. One theory of the origin of eukaryotic cells - those with a true nucleus and other organelles - is that the mitochondria (that release energy from food) and chloroplasts (which makes food through photosynthesis) were at one time, free living prokaryotic organisms. The presence in both of tiny amounts of DNA - genes - supports this idea. The development of eukaryotic cells and then more complex multicellular organisms took another two billion years after the first cells were thought to have formed.

Origin of the First Cell

If life is viewed from the point of view of replicators, that is DNA molecules in the organism, cells satisfy two fundamental conditions: protection from the outside environment and confinement of biochemical activity. The former condition is needed to maintain the fragile DNA chains stable in a varying and sometimes aggressive environment, and may have been the main reason for which cells evolved. The latter is fundamental for the evolution of biological complexity. If freely-floating DNA molecules that code for enzymes are not enclosed into cells, the enzymes that benefit a given DNA molecule (for example, by producing nucleotides) will automatically benefit the neighbouring DNA molecules. This might be viewed as "parasitism by default." Therefore the selection pressure on DNA molecules will be much lower, since there is not a definitive advantage for the "lucky" DNA molecule that produces the better enzyme over the others: All molecules in a given neighbourhood are almost equally advantaged.

If all the DNA molecule is enclosed in a cell, then the enzymes coded from the molecule will be kept close to the DNA molecule itself. The DNA molecule will directly enjoy the benefits of the enzymes it codes, and not of others. This means other DNA molecules won't benefit from a positive mutation in a neighbouring molecule: this in turn means that positive mutations give immediate and selective advantage to the replicator bearing it, and not on others. This is thought to have been the one of the main driving force of evolution of life as we know it. (Note. This is more a metaphor given for simplicity than complete accuracy since the earliest molecules of life, probably up to the stage of cellular life, were most likely RNA molecules that acted as both replicators and enzymes: see RNA world hypothesis. However, the core of the reasoning is the same.)

Biochemically, cell-like spheroids formed by proteinoids are observed by heating amino acids with phosphoric acid as a catalyst. They bear much of the basic features provided by cell membranes. Proteinoid-based protocells enclosing RNA molecules could (but not necessarily should) have been the first cellular life forms on Earth.

Another theory holds that the turbulent shores of the ancient coastal waters may have served as a mammoth laboratory, aiding in the countless experiments necessary to bring about the first cell. Waves breaking on the shore create a delicate foam composed of bubbles. Winds sweeping across the ocean have a tendency to drive things to shore, much like driftwood collecting on the beach. It is possible that organic molecules were concentrated on the shorelines in much the same way. Shallow coastal waters also tend to be warmer, further concentrating the molecules through evapouration. While bubbles comprised of mostly water tend to burst quickly, oily bubbles happen to be much more stable, lending more time to the particular bubble to perform these crucial experiments. The Phospholipid is a good example of a common oily compound prevalent in the prebiotic seas. Phospholipids can be constructed in one's mind as a hydrophilic head on one end, and a hydrophobic tail on the other. Phospholipids also possess an important characteristic, that is having the function to link together to form a bilayer membrane. A lipid monolayer bubble can only contain oil, and is therefore not conducive to harbouring water-soluble organic molecules. On the other hand, a lipid bilayer bubble can contain water, and was a likely precursor to the modern cell membrane. If a protein came along that increased the integrity of its parent bubble, then that bubble had an advantage, and was placed at the top of the natural selection waiting list. Primitive reproduction can be envisioned when the bubbles burst, releasing the results of the experiment into the surrounding medium. Once enough of the 'right stuff' was released into the medium, the development of the first prokaryotes, eukaryotes, and multi-cellular organisms could be achieved.

Origin of Eukaryotic Cells

The eukaryotic cell seems to have evolved from a symbiotic community of prokaryotic cells. It is almost certain that DNA-bearing organelles like the mitochondria and the chloroplasts are what remains of ancient symbiotic oxygen-breathing bacteria and cyanobacteria, respectively, where the rest of the cell seems to be derived from an ancestral archaean prokaryote cell – a theory termed the endosymbiotic theory.

There is still considerable debate about whether organelles like the hydrogenosome predated the origin of mitochondria, or viceversa: see the hydrogen hypothesis for the origin of eukaryotic cells.

Properties of Cells

Each cell is at least somewhat self-contained and self-maintaining: it can take in nutrients, convert these nutrients into energy, carry out specialized functions, and reproduce as necessary. Each cell stores its own set of instructions for carrying out each of these activities.

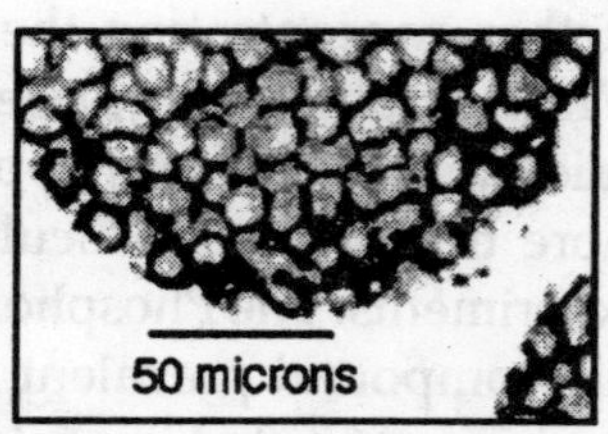

Fig. Cell

Mouse cells grown in a culture dish. These cells grow in large clumps, but each individual cell is about 10 micrometres across.

All cells share several abilities:

Reproduction by cell division (binary fission, mitosis or meiosis).

Use of enzymes and other proteins coded for by DNA genes and made via messenger RNA intermediates and ribosomes.

Metabolism, including taking in raw materials, building cell components, converting energy, molecules and releasing by-products. The functioning of a cell depends upon its ability to extract and use chemical energy stored in organic molecules. This energy is derived from metabolic pathways.

Response to external and internal stimuli such as changes in temperature, pH or nutrient levels.

Cell contents are contained within a cell surface membrane that contains proteins and a lipid bilayer.

Some prokaryotic cells contain important internal membrane-bound compartments, but eukaryotic cells have a highly specialized endomembrane system characterized by regulated traffic and transport of vesicles.

Anatomy of Cells

Basic Cellular Components

All cells contain cytoplasm, plasma membrane, and DNA.

Cytoplasm is the viscous contents of a cell, including proteins, ribosomes, metabolites and ions. Ribosomes are the sites of protein synthesis.

Plasma membrane is the cell membrane surrounding cytoplasm. It consists of phospholipid bilayer, associated proteins and carbohydrates. Phospholipid bilayer is also the basic constituent of other biomembranes.

DNA (deoxyribonucleic acid) is the genetic material. An eukaryotic cell contains several DNA molecules, located in the nucleus and mitochondria which are membrane-bound organelles. A prokaryotic cell contains a single DNA molecule, which has no specific boundary with the cytoplasm.

Classification of Cells and Organisms

All cells are divided into two types: prokaryotic cells and eukaryotic cells.

The prokaryotic cell does not have a nucleus.

The eukaryotic cell contains a nucleus.

Eukaryotes are the organisms made up of eukaryotic cells. They include protista, fungi, animals and plants. Prokaryotes include archaebacteria and eubacteria. They are single-cell organisms.

More recently, "archaebacteria" have been placed in a category outside "bacteria", because they are quite different from the ordinary bacteria. According to the new classification, prokaryotes are divided into archaea and bacteria, where "archaea" is equivalent to "archaebacteria", and "bacteria" is the same as "eubacteria".

Archaea live in extreme environments. They may be organized into three groups: Methanogens live in anaerobic environment such as swamps. They produce methane and cannot tolerate exposure to oxygen.

Extreme halophiles live in very high concentrations of salt (NaCl), e.g., the Dead Sea and the Great Salt Lake.

Extreme thermophiles live in hot, sulphur rich and low pH environment, such as hot springs, geysers and fumaroles in the Yellowstone National Park

Two types of major cells are eukaryotic and prokaryotic. Prokaryotic cells are usually singletons, while eukaryotic cells are usually found in multicellular organisms.

Prokaryotic Cells

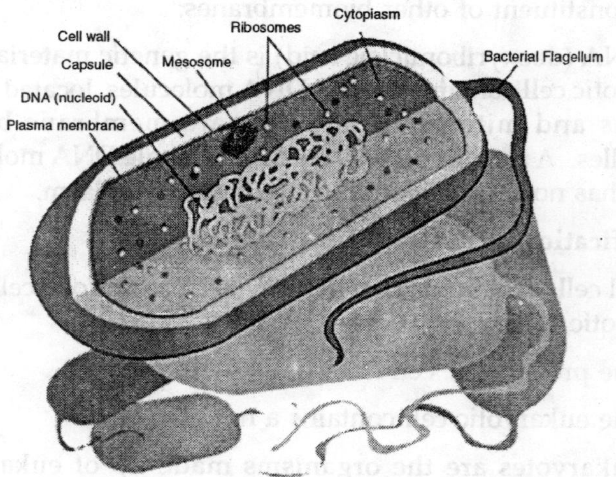

Fig. Diagram of a Typical Prokaryotic Cell

Prokaryotes are distinguished from eukaryotes on the basis of nuclear organization, specifically their lack of a nuclear membrane. Prokaryotes also lack most of the intracellular organelles and structures that are characteristic of eukaryotic cells (an important exception is the ribosomes, which are present in both prokaryotic and eukaryotic cells). Most of the functions of organelles, such as mitochondria, chloroplasts, and the Golgi apparatus, are taken over by the prokaryotic plasma membrane. Prokaryotic cells have three architectural regions: appendages called flagella and pili – proteins attached to the cell surface; a cell envelope consisting of a capsule, a cell wall, and a plasma membrane; and a cytoplasmic region that contains the cell genome (DNA) and ribosomes and various sorts of inclusions. Other differences include:

The plasma membrane (a phospholipid bilayer) separates the interior of the cell from its environment and serves as a filter and communications beacon.

Most prokaryotes have a cell wall (some exceptions are Mycoplasma (a bacterium) and Thermoplasma (an archaeon)). It consists of peptidoglycan in bacteria, and acts as an additional barrier against exterior forces. It also prevents the cell from "exploding" (cytolysis) from osmotic pressure against a hypotonic environment. A cell wall is also present in some eukaryotes like fungi, but has a different chemical composition.

A prokaryotic chromosome is usually a circular molecule (an exception is that of the bacterium Borrelia burgdorferi, which causes Lyme disease). Even without a real nucleus, the DNA is condensed in a nucleoid. Prokaryotes can carry extrachromosomal DNA elements called plasmids, which are usually circular. Plasmids can carry additional functions, such as antibiotic resistance.

Eukaryotic Cells

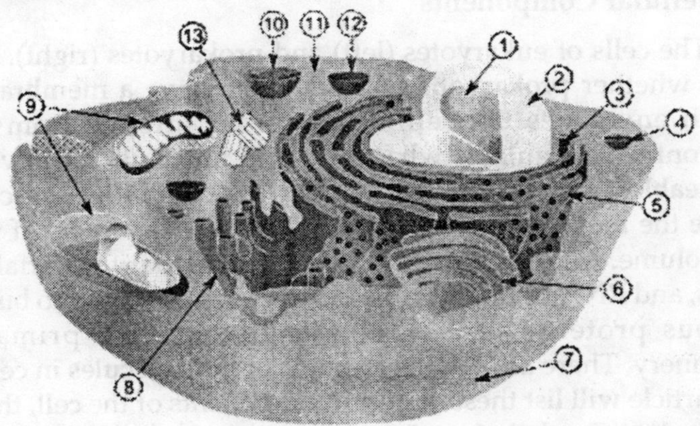

Fig. Diagram of a typical eukaryotic cell, showing subcellular components. Organelles: (1) nucleolus (2) nucleus (3) ribosome (4) vesicle (5) rough endoplasmic reticulum (ER) (6) Golgi apparatus (7) Cytoskeleton (8) smooth ER (9) mitochondria (10) vacuole (11) cytoplasm (12) lysosome (13) centrioles

Eukaryotic cells are about 10 times the size of a typical prokaryote and can be as much as 1000 times greater in volume. The major difference between prokaryotes and eukaryotes is that eukaryotic cells contain membrane-bound compartments in which specific metabolic activities take place. Most important among these is the presence of a cell nucleus, a membrane-delineated compartment that houses the eukaryotic cell's DNA. It is this nucleus that gives the eukaryote its name, which means "true nucleus". Other differences include:

The plasma membrane resembles that of prokaryotes in function, with minor differences in the setup. Cell walls may or may not be present.

The eukaryotic DNA is organized in one or more linear molecules, called chromosomes, which are associated with histone proteins. All chromosomal DNA is stored in the cell nucleus, separated from the cytoplasm by a membrane. Some eukaryotic organelles also contain some DNA.

Eukaryotes can move using cilia or flagella. The flagella are more complex than those of prokaryotes.

Subcellular Components

The cells of eukaryotes (left) and prokaryotes (right). All cells, whether prokaryotic or eukaryotic, have a membrane, which envelopes the cell, separates its interior from its environment, regulates what moves in and out (selectively permeable), and maintains the electric potential of the cell. Inside the membrane, a salty cytoplasm takes up most of the cell volume. All cells possess DNA, the hereditary material of genes, and RNA, containing the information necessary to build various proteins such as enzymes, the cell's primary machinery. There are also other kinds of biomolecules in cells. This article will list these primary components of the cell, then briefly describe their function.

Cell Membrane: A Cell's Defining Boundary

The cytoplasm of a cell is surrounded by a plasma membrane. The plasma membrane in plants and prokaryotes is usually covered by a cell wall. This membrane serves to separate and protect a cell from its surrounding environment and is made mostly from a double layer of lipids (hydrophobic fat-like molecules) and hydrophilic phosphorus molecules. Hence, the layer is called a phospholipid bilayer. It may also be called a fluid mosaic membrane. Embedded within this membrane is a variety of protein molecules that act as channels and pumps that move different molecules into and out of the cell. The membrane is said to be 'semi-permeable', in that it can either let a substance (molecule or ion) pass through freely, pass through to a limited extent or not pass through at all. Cell surface membranes also contain receptor proteins that allow cells to detect external signalling molecules such as hormones.

Cytoskeleton: A Sell's Scaffold

The cytoskeleton acts to organize and maintain the cell's shape; anchors organelles in place; helps during endocytosis, the uptake of external materials by a cell, and cytokinesis, the separation of daughter cells after cell division; and moves parts of the cell in processes of growth and mobility. Eukaryotic cytoskeleton is composed of microfilaments, intermediate filaments and microtubules. There is a great number of proteins associated with them, each controlling a cell's structure by directing, bundling, and aligning filaments.

Genetic Material

Two different kinds of genetic material exist: deoxyribonucleic acid (DNA) and ribonucleic acid (RNA). Most organisms use DNA for their long-term information storage, but some viruses (e.g., retroviruses) have RNA as their genetic material. The biological information contained in an organism is encoded in its DNA or RNA sequence. RNA is also used for information transport (e.g., mRNA) and enzymatic functions (e.g., ribosomal RNA) in organisms that use DNA for the genetic code itself.

Prokaryotic genetic material is organized in a simple circular DNA molecule (the bacterial chromosome) in the nucleoid region of the cytoplasm. Eukaryotic genetic material is divided into different, linear molecules called chromosomes inside a discrete nucleus, usually with additional genetic material in some organelles like mitochondria and chloroplasts (see endosymbiotic theory).

A human cell has genetic material in the nucleus (the nuclear genome) and in the mitochondria (the mitochondrial genome). In humans the nuclear genome is divided into 46 linear DNA molecules called chromosomes. The mitochondrial genome is a circular DNA molecule separate from the nuclear DNA. Although the mitochondrial genome is very small, it codes for some important proteins.

Foreign genetic material (most commonly DNA) can also be artificially introduced into the cell by a process called transfection. This can be transient, if the DNA is not inserted into the cell's genome, or stable, if it is.

Organelles

The human body contains many different organs, such as the heart, lung, and kidney, with each organ performing a different function. Cells also have a set of "little organs," called organelles, that are adapted and/or specialized for carrying out one or more vital functions. Membrane-bound organelles are found only in eukaryotes.

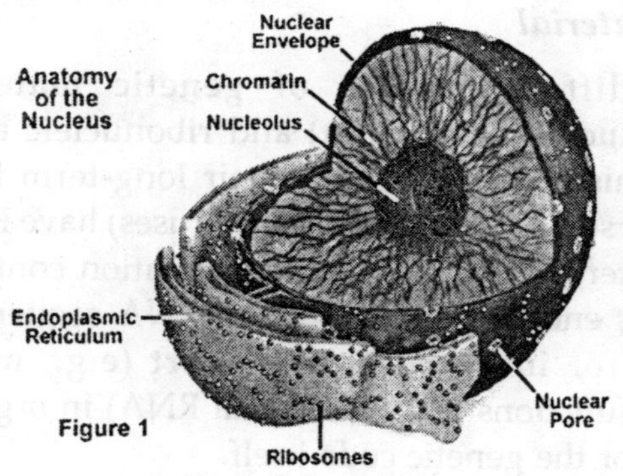

Fig. Cell Nucleus

The cell nucleus is the most conspicuous organelle found in a eukaryotic cell. It houses the cell's chromosomes, and is the place where almost all DNA replication and RNA synthesis occur. The nucleus is spheroid in shape and separated from the cytoplasm by a double membrane called the nuclear envelope. The nuclear envelope isolates and protects a cell's DNA from various molecules that could accidentally damage its structure or interfere with its processing. During processing, DNA is transcribed, or copied into a special RNA, called mRNA. This mRNA is then transported out of the nucleus, where it is translated into a specific protein molecule. In prokaryotes, DNA processing takes place in the cytoplasm.,

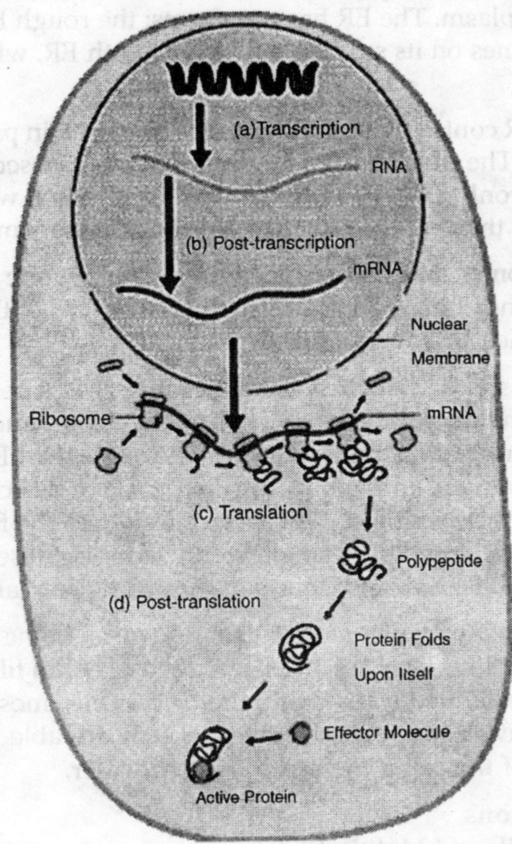

Fig. Mitochondria and Chloroplasts (the power generators)

Mitochondria are self-replicating organelles that occur in various numbers, shapes, and sizes in the cytoplasm of all eukaryotic cells. As mitochondria contain their own genome that is separate and distinct from the nuclear genome of a cell, they play a critical role in generating energy in the eukaryotic cell, organelles that are modified chloroplasts; they are broadly called plastids, and are often involved in storage.

Endoplasmic reticulum and Golgi apparatus (macromolecule managers)

The endoplasmic reticulum (ER) is the transport network for molecules targeted for certain modifications and specific destinations, as compared to molecules that will float freely in the cytoplasm. The ER has two forms: the rough ER, which has ribosomes on its surface, and the smooth ER, which lacks them.

The ER contains many Ribosomes (the protein production machine) : The ribosome is a large complex composed of many molecules, only exist floating freely in the cytosol, whereas in eukaryotes they can be either free or bound to membranes.

Lysosomes and Peroxisomes (of the eukaryotic cell): The cell could not house such destructive enzymes if they were not contained in a membrane-bound system.

Centrosome (the cytoskeleton organiser): The centrosome produces the microtubules of a cell - a key component of the cytoskeleton. It directs the transport through the ER and the Golgi apparatus. Centrosomes are composed of two centrioles, which separate during cell division and help in the formation of the mitotic spindle. A single centrosome is present in the animal cells. They are also found in some fungi and algae cells.

Vacuoles: Vacuoles store food and waste. Some vacuoles store extra water. They are often described as liquid filled space and are surrounded by a membrane. Some cells, most notably Amoeba have contractile vacuoles, which are able to pump water out of the cell if there is too much water.

Cell Functions

Cell Growth and Metabolism

Between successive cell divisions, cells grow through the functioning of cellular metabolism.

Cell metabolism is the process by which individual cells process nutrient molecules. Metabolism has two distinct divisions: catabolism, in which the cell breaks down complex molecules to produce energy and reducing power, and anabolism, wherein the cell uses energy and reducing power to construct complex molecules and perform other biological functions. Complex sugars consumed by the organism can be broken down into a less chemically-complex sugar molecule called glucose. Once inside the cell, glucose is broken down to make adenosine triphosphate (ATP), a form of energy, via two different pathways.

The first pathway, glycolysis, requires no oxygen and is referred to as anaerobic metabolism. Each reaction is designed to produce some hydrogen ions that can then be used to make energy packets (ATP). In prokaryotes, glycolysis is the only method used for converting energy.

The second pathway, called the Krebs cycle, or citric acid cycle, occurs inside the mitochondria and is capable of generating enough ATP to run all the cell functions.

An overview of protein synthesis: Within the nucleus of the cell (light blue), genes (DNA, dark blue) are transcribed into RNA. This RNA is then subject to post-transcriptional modification and control, resulting in a mature mRNA (red) that is then transported out of the nucleus and into the cytoplasm (peach), where it undergoes translation into a protein. mRNA is translated by ribosomes (purple) that match the three-base codons of the mRNA to the three-base anti-codons of the appropriate tRNA. Newly-synthesized proteins (black) are often further modified, such as by binding to an effector molecule (orange), to become fully active.

Creation of New Cells

Cell division involves a single cell (called a mother cell) dividing into two daughter cells. This leads to growth in multicellular organisms (the growth of tissue) and to procreation (vegetative reproduction) in unicellular organisms.

Prokaryotic cells divide by binary fission. Eukaryotic cells usually undergo a process of nuclear division, called mitosis, followed by division of the cell, called cytokinesis. A diploid cell may also undergo meiosis to produce haploid cells, usually four. Haploid cells serve as gametes in multicellular organisms, fusing to form new diploid cells.

DNA replication, or the process of duplicating a cell's genome, is required every time a cell divides. Replication, like all cellular activities, requires specialized proteins for carrying out the job.

Protein Synthesis

Cells are capable of synthesizing new proteins, which are essential for the modulation and maintenance of cellular activities. This process involves the formation of new protein molecules from amino acid building blocks based on information encoded in DNA/RNA. Protein synthesis generally consists of two major steps: transcription and translation.

Transcription is the process where genetic information in DNA is used to produce a complimentary RNA strand. This RNA strand is then processed to give messenger RNA (mRNA), which is free to migrate through the cell. mRNA molecules bind to protein-RNA complexes called ribosomes located in the cytosol, where they are translated into polypeptide sequences. The ribosome mediates the formation of a polypeptide sequence based on the mRNA sequence. The mRNA sequence directly relates to the polypeptide sequence by binding to transfer RNA (tRNA) adapter molecules in binding pockets within the ribosome. The new polypeptide then folds into a functional 3D protein molecule.

CELLS: CELLULAR ORGANIZATION

According to the Cell Theory, all living things are composed of one or more cells. Cells fall into prokaryotic and eukaryotic types. Prokaryotic cells are smaller (as a general rule) and lack much of the internal compartmentalization and complexity of eukaryotic cells. No matter which type of cell we are considering, all cells have certain features in common: cell membrane, DNA, cytoplasm, and ribosomes.

CELL SIZE AND SHAPE

The shapes of cells are quite varied with some, such as neurons, being longer than they are wide and others, such as parenchyma (a common type of plant cell) and erythrocytes (red blood cells) being equidimensional. Some cells are encased in a rigid wall, which constrains their shape, while others have a flexible cell membrane (and no rigid cell wall).

The size of cells is also related to their functions. Eggs (or to use the latin word, ova) are very large, often being the largest cells an organism produces. The large size of many eggs is related to the process of development that occurs after the egg is fertilized, when the contents of the egg (now termed a zygote) are used in a rapid series of cellular divisions, each requiring tremendous amounts of energy that is available in the zygote cells. Later in life the energy must be acquired, but at first a sort of inheritance/trust fund of energy is used.

The Cell Membrane

The cell membrane functions as a semi-permeable barrier, allowing a very few molecules across it while fencing the majority of organically produced chemicals inside the cell. Electron microscopic examinations of cell membranes have led to the development of the lipid bilayer model (also referred to as the fluid-mosaic model). The most common molecule in the model is the phospholipid, which has a polar (hydrophilic) head and two nonpolar (hydrophobic) tails. These phospholipids are aligned tail to tail so the nonpolar areas form a hydrophobic region between the hydrophilic heads on the inner and outer surfaces of the membrane. This layering is termed a bilayer since an electron microscopic technique known as freeze-fracturing is able to split the bilayer.

Cholesterol is another important component of cell membranes embedded in the hydrophobic areas of the inner (tail-tail) region. Most bacterial cell membranes do not contain cholesterol.

Proteins are suspended in the inner layer, although the more hydrophilic areas of these proteins "stick out" into the cells interior and outside of the cell. These proteins function as gateways that will, in exchange for a price, allow certain molecules to cross into and out of the cell. These integral proteins are sometimes known as gateway proteins. The outer surface of the membrane will tend to be rich in glycolipids, which have their hydrophobic tails embedded in the hydrophobic region of the membrane and their heads exposed outside the cell. These, along with carbohydrates attached to the integral proteins, are thought to function in the recognition of self.

The contents (both chemical and organelles)of the cell are termed protoplasm, and are further subdivided into cytoplasm (all of the protoplasm except the contents of the nucleus) and nucleoplasm (all of the material, plasma and DNA etc. within the nucleus).

The Cell Wall

Not all living things have cell walls, most notably animals and many of the more animal-like Protistans. Bacteria have cell walls containing peptidoglycan. Plant cells have a variety of chemicals incorporated in their cell walls. Cellulose is the most common chemical in the plant primary cell wall. Some plant cells also have lignin and other chemicals embedded in their secondary walls. The cell wall is located outside the plasma membrane. Plasmodesmata are connections through which cells communicate chemically with each other through their thick walls. Fungi and many protists have cell walls although they do not contain cellulose, rather a variety of chemicals (chitin for fungi).

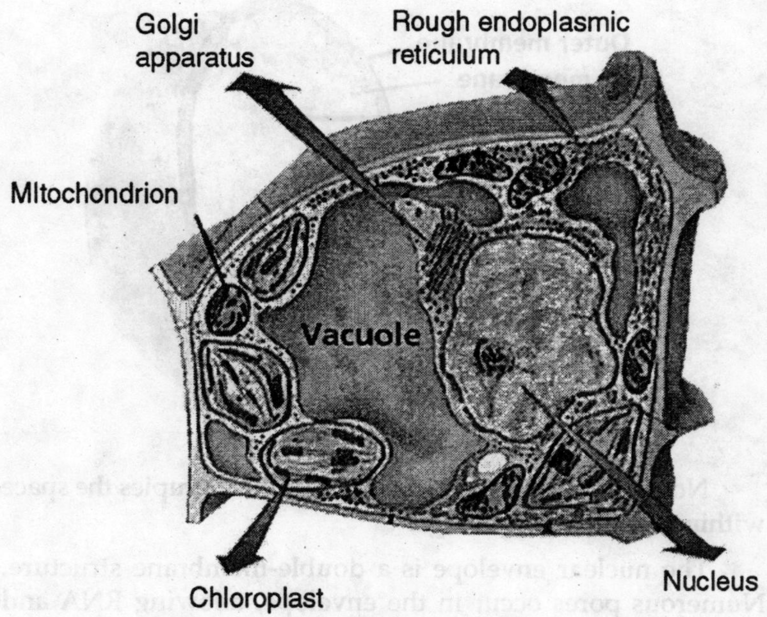

Fig. Structure of a Typical Plant Cell

The Nucleus

The nucleus occurs only in eukaryotic cells, and is the location of the majority of different types of nucleic acids. Van Hammerling's experiment showed the role of the nucleus in controlling the shape and features of the cell.

Deoxyribonucleic acid, DNA, is the physical carrier of inheritance and with the exception of plastid DNA (cpDNA and mDNA, see below) all DNA is restricted to the nucleus. Ribonucleic acid, RNA, is formed in the nucleus by coding off of the DNA bases.

RNA moves out into the cytoplasm. The nucleolus is an area of the nucleus (usually 2 nucleoli per nucleus) where ribosomes are constructed.

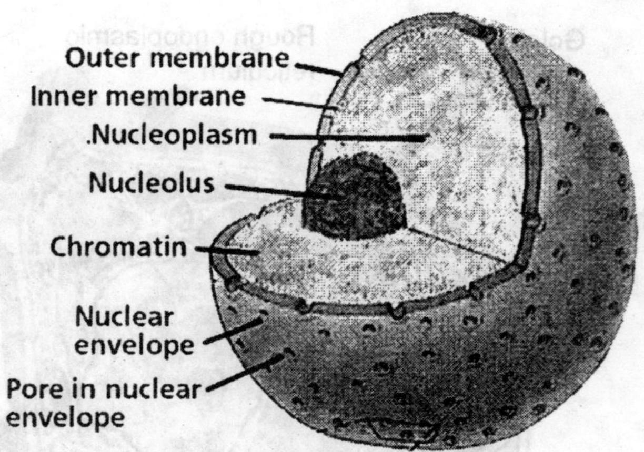

Fig. Structure of the Nucleus

Note the chromatin, uncoiled DNA that occupies the space within the nuclear envelope.

The nuclear envelope is a double-membrane structure. Numerous pores occur in the envelope, allowing RNA and other chemicals to pass, but the DNA not to pass.

Structure of the nuclear envelope and nuclear pores.

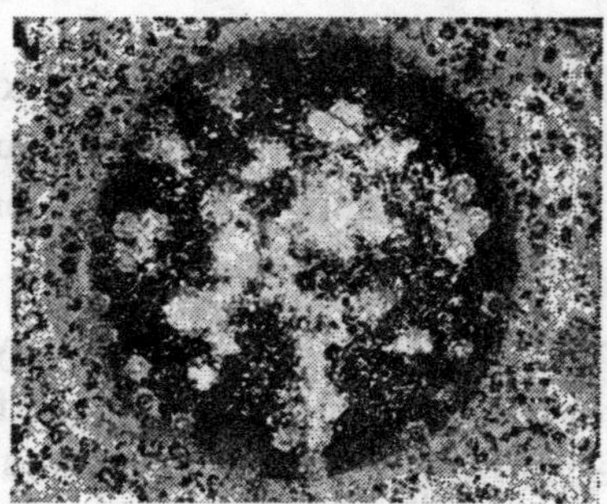

Fig. Nucleus with Nuclear Pores

Cytoplasm

The cytoplasm was defined earlier as the material between the plasma membrane (cell membrane) and the nuclear envelope. Fibrous proteins that occur in the cytoplasm, referred to as the cytoskeleton maintain the shape of the cell as well as anchoring organelles, moving the cell and controlling internal movement of structures. Microtubules function in cell division and serve as a "temporary scaffolding" for other organelles. Actin filaments are thin threads that function in cell division and cell motility. Intermediate filaments are between the size of the microtubules and the actin filaments.

Vacuoles and vesicles

Vacuoles are single-membrane organelles that are essentially part of the outside that is located within the cell. The single membrane is known in plant cells as a tonoplast. Many organisms will use vacuoles as storage areas. Vesicles; are much smaller than vacuoles and function in transport within and to the outside of the cell.

Ribosomes

Ribosomes are the sites of protein synthesis. They are not membrane-bound and thus occur in both prokaryotes and eukaryotes. Eukaryotic ribosomes are slightly larger than prokaryotic ones. Structurally the ribosome consists of a small and larger subunit. Biochemically the ribosome consists of ribosomal RNA (rRNA) and some 50 structural proteins. Often ribosomes cluster on the endoplasmic reticulum, in which case they resemble a series of factories adjoining a railroad line.

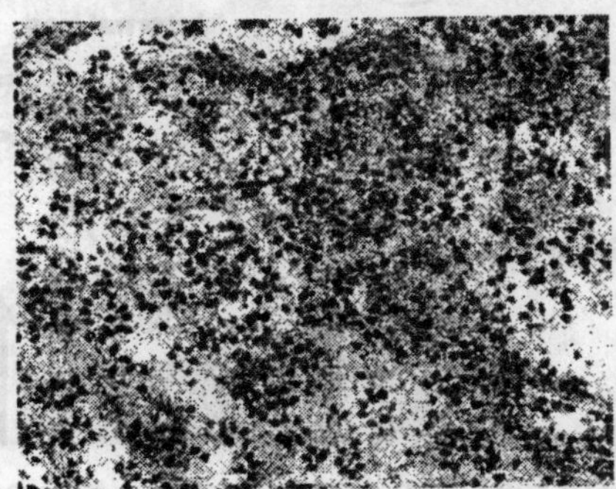

Fig. Ribosomes and Polyribosomes - Liver Cell

Endoplasmic reticulum

Endoplasmic reticulum is a mesh of interconnected membranes that serve a function involving protein synthesis and transport. Rough endoplasmic reticulum (Rough ER) is so-named because of its rough appearance due to the numerous ribosomes that occur along the ER. Rough ER connects to the nuclear envelope through which the messenger RNA (mRNA) that is the blueprint for proteins travels to the ribosomes. Smooth ER; lacks the ribosomes characteristic of Rough ER and is thought to be involved in transport and a variety of other functions.

The endoplasmic reticulum. Rough endoplasmic reticulum is on the left, smooth endoplasmic reticulum is on the right.

Rough Endoplasmic Reticulum with Ribosomes:

Golgi Apparatus and Dictyosomes

Golgi Complexes are flattened stacks of membrane-bound sacs. They function as a packaging plant, modifying vesicles from the Rough ER. New membrane material is assembled in various cisternae of the golgi.

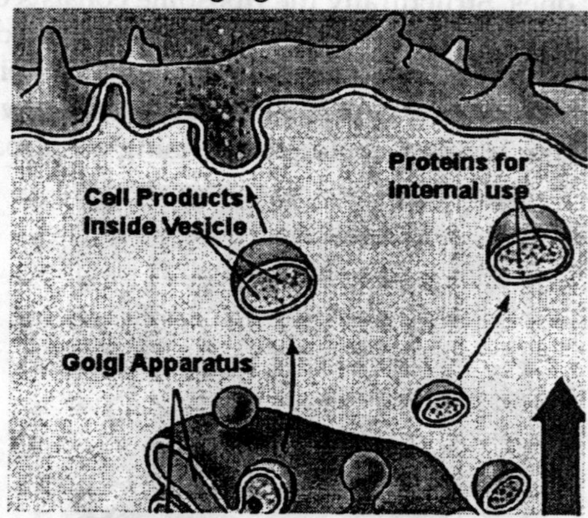

Fig. Structure of the Golgi Apparatus and its Functioning in Vesicle-mediated Transport

Lysosomes: Lysosomes are relatively large vesicles formed by the Golgi. They contain hydrolytic enzymes that could destroy the cell. Lysosome contents function in the extracellular breakdown of materials.

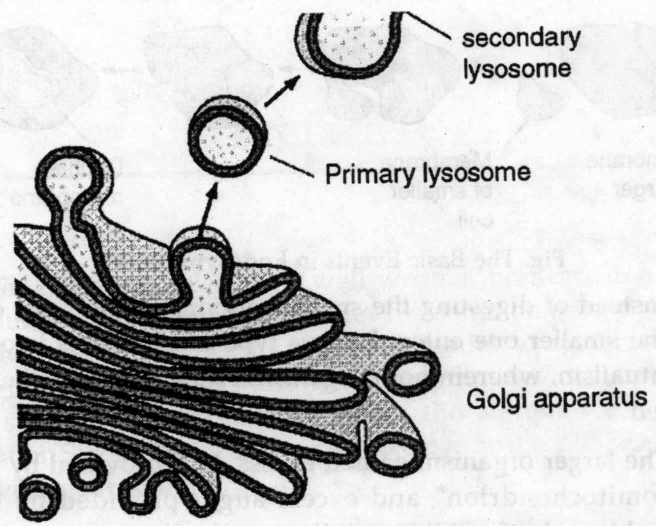

Fig. Role of the Golgi in Forming Lysosomes

Mitochondria

Mitochondria contain their own DNA (termed mDNA) and are thought to represent bacteria-like organisms incorporated into eukaryotic cells over 700 million years ago (perhaps even as far back as 1.5 billion years ago).

They function as the sites of energy release (following glycolysis in the cytoplasm) and ATP formation (by chemiosmosis). The mitochondrion has been termed the powerhouse of the cell. Mitochondria are bounded by two membranes. The inner membrane folds into a series of cristae, which are the surfaces on which ATP is generated.

Mitochondria and endosymbiosis: During the 1980s, Lynn Margulis proposed the theory of endosymbiosis to explain the origin of mitochondria and chloroplasts from permanent resident prokaryotes. According to this idea, a larger prokaryote (or perhaps early eukaryote) engulfed or surrounded a smaller prokaryote some 1.5 billion to 700 million years ago.

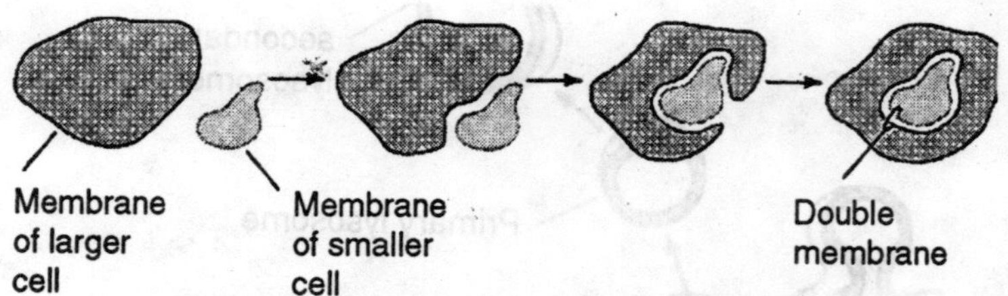

Fig. The Basic Events in Endosymbiosis

Instead of digesting the smaller organisms the large one and the smaller one entered into a type of symbiosis known as mutualism, wherein both organisms benefit and neither is harmed.

The larger organism gained excess ATP provided by the "protomitochondrion" and excess sugar provided by the "protochloroplast", while providing a stable environment and the raw materials the endosymbionts required. This is so strong that now eukaryotic cells cannot survive without mitochondria (likewise photosynthetic eukaryotes cannot survive without chloroplasts), and the endosymbionts can not survive outside their hosts. Nearly all eukaryotes have mitochondria. Mitochondrial division is remarkably similar to the prokaryotic methods that will be studied later in this course.

Plastids

Plastids are also membrane-bound organelles that only occur in plants and photosynthetic eukaryotes.

Chloroplasts are the sites of photosynthesis in eukaryotes. They contain chlorophyll, the green pigment necessary for photosynthesis to occur, and associated accessory pigments (carotenes and xanthophylls) in photosystems embedded in membranous sacs, thylakoids (collectively a stack of thylakoids are a granum [plural = grana]) floating in a fluid termed the stroma. Chloroplasts contain many different types of accessory pigments, depending on the taxonomic group of the organism being observed.

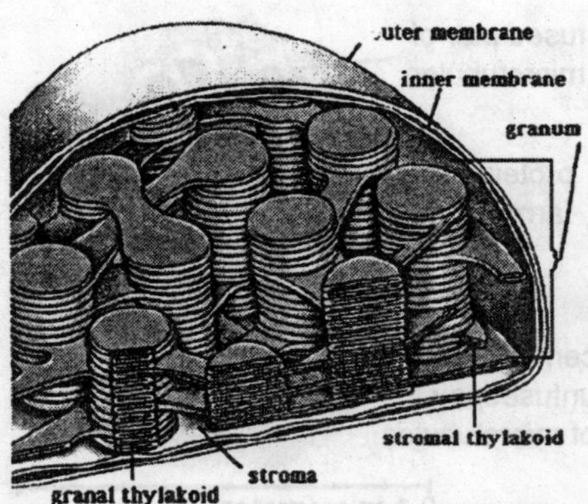

Figure: Structure of the Chloroplast

Chloroplasts and endosymbiosis: Like mitochondria, chloroplasts have their own DNA, termed cpDNA. Chloroplasts of Green Algae (Protista) and Plants (descendants of some Green Algae) are thought to have originated by endosymbiosis of a prokaryotic alga similar to living Prochloron (Prochlorobacteria). Chloroplasts of Red Algae (Protista) are very similar biochemically to cyanobacteria (also known as blue-green bacteria [algae to chronologically enhanced folks like myself:)]). Endosymbiosis is also invoked for this similarity, perhaps indicating more than one endosymbiotic event occurred.

Leukoplasts store starch, sometimes protein or oils.

Chromoplasts store pigments associated with the bright colours of flowers and/or fruits.

Cell Movement

Cell movement is both internal, referred to as cytoplasmic streaming and external, referred to as motility. Internal movements of organelles are governed by actin filaments. These filaments make an area in which organelles such as chloroplasts can move. Internal movement is known as cytoplasmic streaming. External movement of cells is determined by special organelles for locomotion.

Cilia and flagella are similar except for length, cilia being much shorter. They both have the characteristic 9 + 2 arrangement of microtubules.

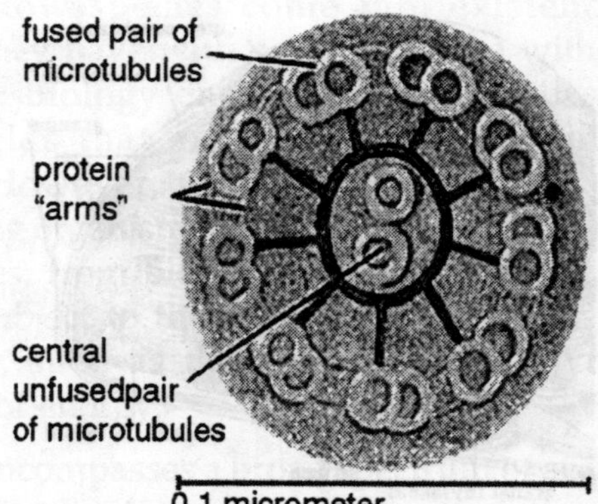

Fig. The 9+2 Arrangement of Microtubules in a Flagellum or Cilium

Flagella work as whips pulling (as in Chlamydomonas or Halosphaera) or pushing (dinoflagellates, a group of single-celled Protista) the organism through the water. Cilia work like oars on a viking longship (Paramecium has 17,000 such oars covering its outer surface).

Pseudopodia are used by many cells, such as Amoeba, Chaos (Pelomyxa) and human leukocytes (white blood cells). These are not structures as such but rather are associated with actin near the moving edge.

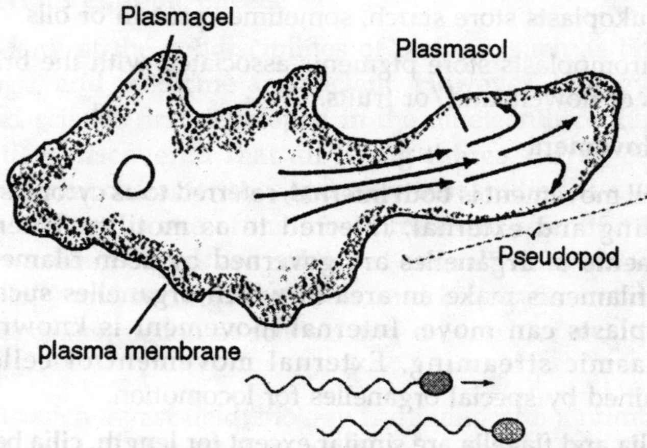

Fig. Formation and Functioning of a Pseudopod by an Amoeboid Cell.

TRANSPORT IN AND OUT OF CELL

Water and Solute Movement

Cell membranes act as barriers to most, but not all, molecules. Development of a cell membrane that could allow some materials to pass while constraining the movement of other molecules was a major step in the evolution of the cell. Cell membranes are differentially (or semi-) permeable barriers separating the inner cellular environment from the outer cellular (or external) environment.

Water potential is the tendency of water to move from an area of higher concentration to one of lower concentration. Energy exists in two forms: potential and kinetic. Water molecules move according to differences in potential energy between where they are and where they are going. Gravity and pressure are two enabling forces for this movement. These forces also operate in the hydrologic (water) cycle. Remember in the hydrologic cycle that water runs downhill (likewise it falls from the sky, to get into the sky it must be acted on by the sun and evapourated, thus needing energy input to power the cycle).

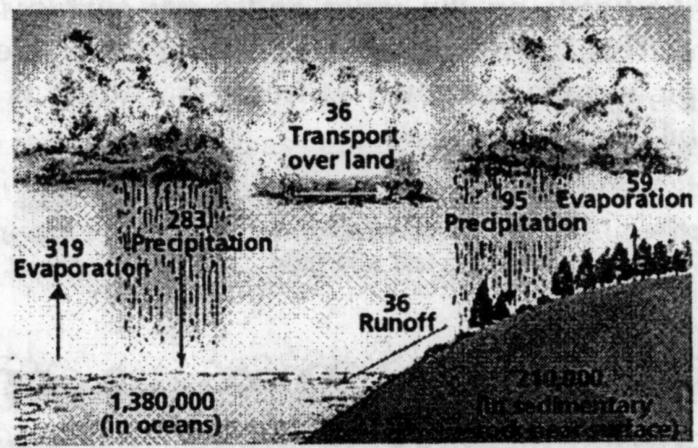

The Hydrologic Cycle: Diffusion is the net movement of a substance (liquid or gas) from an area of higher concentration to one of lower concentration. Nearer the source the concentration of a given substance increases. You probably experience this in class when someone arrives freshly doused in perfume or cologne, especially the cheap stuff.

Since the molecules of any substance (solid, liquid, or gas) are in motion when that substance is above absolute zero (0 degrees Kelvin or -273 degrees C), energy is available for movement of the molecules from a higher potential state to a lower potential state, just as in the case of the water discussed above. The majority of the molecules move from higher to lower concentration, although there will be some that move from low to high. The overall (or net) movement is thus from high to low concentration. Eventually, if no energy is input into the system the molecules will reach a state of equilibrium where they will be distributed equally throughout the system.

The Cell Membrane: The cell membrane functions as a semi-permeable barrier, allowing a very few molecules across it while fencing the majority of organically produced chemicals inside the cell. Electron microscopic examinations of cell membranes have led to the development of the lipid bilayer model (also referred to as the fluid-mosaic model). The most common molecule in the model is the phospholipid, which has a polar (hydrophilic) head and two nonpolar (hydrophobic) tails. These phospholipids are aligned tail to tail so the nonpolar areas form a hydrophobic region between the hydrophilic heads on the inner and outer surfaces of the membrane. This layering is termed a bilayer since an electron microscopic technique known as freeze-fracturing is able to split the bilayer.

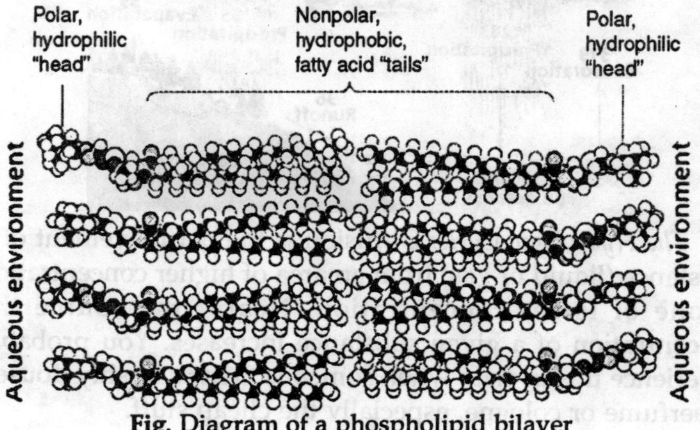

Fig. Diagram of a phospholipid bilayer

Phospholipids and glycolipids are important structural components of cell membranes. Phospholipids are modified so that a phosphate group (PO4-) replaces one of the three fatty acids normally found on a lipid. The addition of this group makes a polar "head" and two nonpolar "tails".

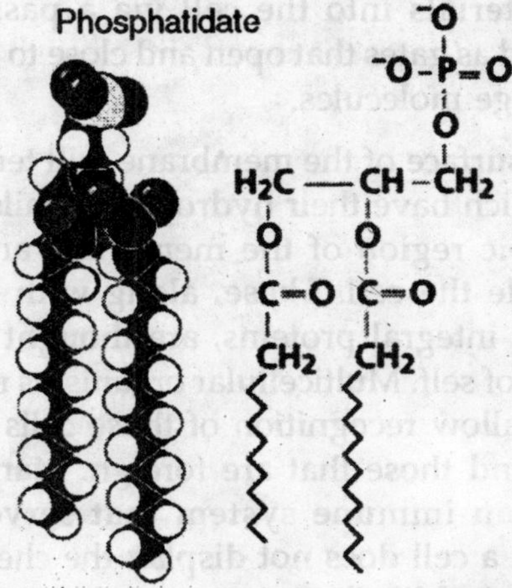

Fig. Structure of a Phospholipid, Space-filling Model (left) and Chain Model (right).

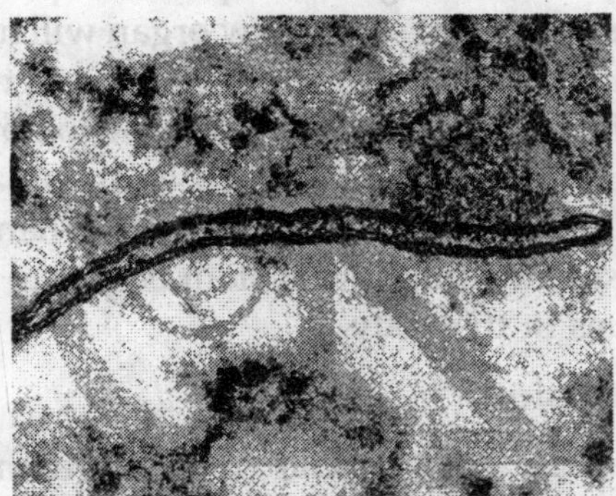

Fig. Cell Membranes from Opposing Neurons

Cholesterol is another important component of cell membranes embedded in the hydrophobic areas of the inner (tail-tail) region. Most bacterial cell membranes do not contain cholesterol.

Proteins are suspended in the inner layer, although the more hydrophilic areas of these proteins "stick out" into the cells interior as well as the outside of the cell. These integral proteins are sometimes known as gateway proteins. Proteins also function in cellular recognition, as binding sites for substances to be brought into the cell, through channels that will allow materials into the cell via a passive transport mechanism, and as gates that open and close to facilitate active transport of large molecules.

The outer surface of the membrane will tend to be rich in glycolipids, which have their hydrophobic tails embedded in the hydrophobic region of the membrane and their heads exposed outside the cell. These, along with carbohydrates attached to the integral proteins, are thought to function in the recognition of self. Multicellular organisms may have some mechanism to allow recognition of those cells that belong to the organism and those that are foreign. Many, but not all, animals have an immune system that serves this sentry function. When a cell does not display the chemical markers that say "Made in Mike", an immune system response may be triggered. This is the basis for immunity, allergies, and autoimmune diseases. Organ transplant recipients must have this response suppressed so the new organ will not be attacked by the immune system, which would cause rejection of the new organ. Allergies are in a sense an over reaction by the immune system. Autoimmune diseases, such as rheumatoid arthritis and systemic lupus erythmatosis, happen when for an as yet unknown reason, the immune system begins to attack certain cells and tissues in the body.

Cells and Diffusion

Water, carbon dioxide, and oxygen are among the few simple molecules that can cross the cell membrane by diffusion (or a type of diffusion known as osmosis). Diffusion is one principle method of movement of substances within cells, as well as the method for essential small molecules to cross the cell membrane. Gas exchange in gills and lungs operates by this process. Carbon dioxide is produced by all cells as a result of cellular metabolic processes. Since the source is inside the cell, the concentration gradient is constantly being replenished/re-elevated, thus the net flow of CO2 is out of the cell. Metabolic processes in animals and plants usually require oxygen, which is in greater concentration inside the cell, thus the net flow of oxygen is into the cell.

Osmosis is the diffusion of water across a semi-permeable (or differentially permeable or selectively permeable) membrane. The cell membrane, along with such things as dialysis tubing and cellulose acetate sausage casing, is such a membrane. The presence of a solute decreases the water potential of a substance. Thus there is more water per unit of volume in a glass of fresh-water than there is in an equivalent volume of sea-water. In a cell, which has so many organelles and other large molecules, the water flow is generally into the cell.

Hypertonic solutions are those in which more solute (and hence lower water potential) is present. Hypotonic solutions are those with less solute (again read as higher water potential). Isotonic solutions have equal (iso-) concentrations of substances. Water potentials are thus equal, although there will still be equal amounts of water movement in and out of the cell, the net flow is zero.

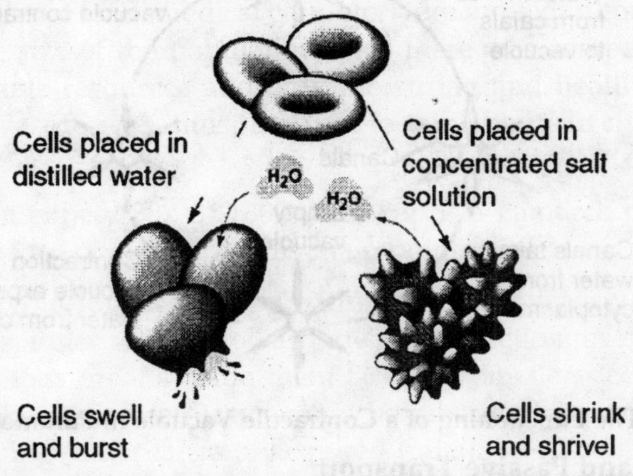

Fig. Water Relations and Cell Shape in Blood Cells

One of the major functions of blood in animals is the maintain an isotonic internal environment. This eliminates the problems associated with water loss or excess water gain in or out of cells. Again we return to homeostasis. Paramecium and other single-celled freshwater organisms have difficulty since they are usually hypertonic relative to their outside environment.

Thus water will tend to flow across the cell membrane, swelling the cell and eventually bursting it. Not good for any cell! The contractile vacuole is the Paramecium's response to this problem. The pumping of water out of the cell by this method requires energy since the water is moving against the concentration gradient.

Since ciliates (and many freshwater protozoans) are hypotonic, removal of water crossing the cell membrane by osmosis is a significant problem. One commonly employed mechanism is a contractile vacuole. Water is collected into the central ring of the vacuole and actively transported from the cell.

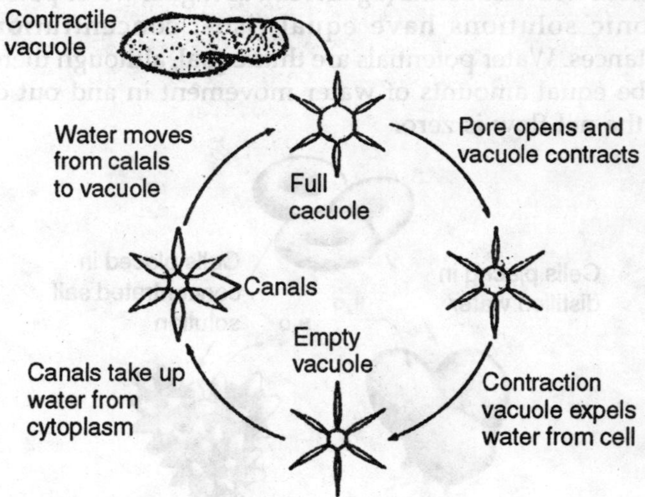

Fig. The Functioning of a Contractile Vacuole in Paramecium

Active and Passive Transport

Passive transport requires no energy from the cell. Examples include the diffusion of oxygen and carbon dioxide, osmosis of water, and facilitated diffusion.

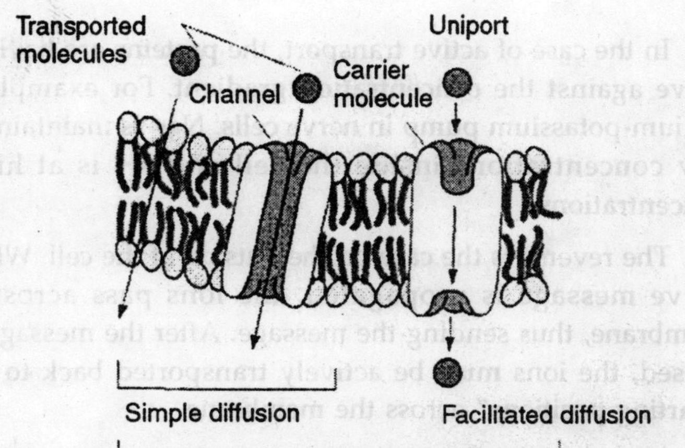

Fig. Types of passive transport

Active transport requires the cell to spend energy, usually in the form of ATP. Examples include transport of large molecules (non-lipid soluble) and the sodium-potassium pump.

Carrier-assisted Transport

The transport proteins integrated into the cell membrane are often highly selective about the chemicals they allow to cross. Some of these proteins can move materials across the membrane only when assisted by the concentration gradient, a type of carrier-assisted transport known as facilitated diffusion.

Both diffusion and facilitated diffusion are driven by the potential energy differences of a concentration gradient. Glucose enters most cells by facilitated diffusion. There seem to be a limiting number of glucose-transporting proteins. The rapid breakdown of glucose in the cell (a process known as glycolysis) maintains the concentration gradient. When the external concentration of glucose increases, however, the glucose transport does not exceed a certain rate, suggesting the limitation on transport.

In the case of active transport, the proteins are having to move against the concentration gradient. For example the sodium-potassium pump in nerve cells. Na+ is maintained at low concentrations inside the cell and K+ is at higher concentrations.

The reverse is the case on the outside of the cell. When a nerve message is propagated, the ions pass across the membrane, thus sending the message. After the message has passed, the ions must be actively transported back to their "starting positions" across the membrane.

This is analogous to setting up 100 dominoes and then tipping over the first one. To reset them you must pick each one up, again at an energy cost. Up to one-third of the ATP used by a resting animal is used to reset the Na-K pump.

Types of Transport Molecules

Uniport transports one solute at a time. Symport transports the solute and a cotransported solute at the same time in the same direction. Antiport transports the solute in (or out) and the co-transported solute the opposite direction. One goes in the other goes out or vice-versa.

Vesicle-mediated transport: Vesicles and vacuoles that fuse with the cell membrane may be utilized to release or transport chemicals out of the cell or to allow them to enter a cell. Exocytosis is the term applied when transport is out of the cell.

Endocytosis is the case when a molecule causes the cell membrane to bulge inward, forming a vesicle. Phagocytosis is the type of endocytosis where an entire cell is engulfed. Pinocytosis is when the external fluid is engulfed. Receptor-mediated endocytosis occurs when the material to be transported binds to certain specific molecules in the membrane. Examples include the transport of insulin and cholesterol into animal cells.

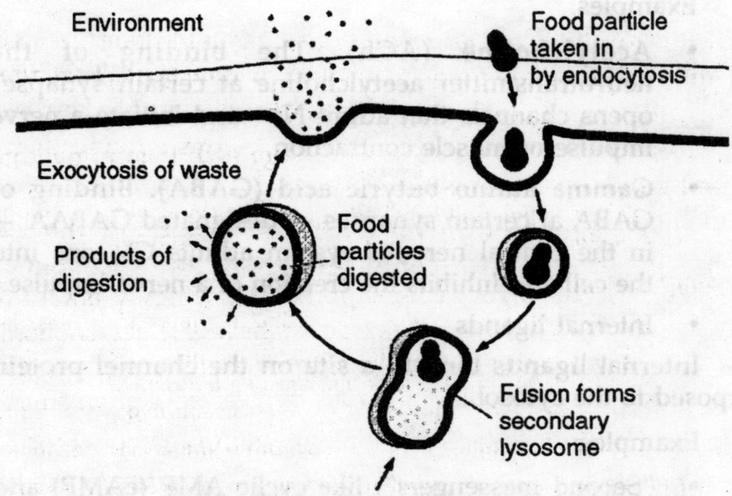

Fig. Endocytosis and Exocytosis

Facilitated Diffusion of Ions

Facilitated diffusion of ions takes place through proteins, or assemblies of proteins, embedded in the plasma membrane. These transmembrane proteins form a water-filled channel through which the ion can pass down its concentration gradient.

The transmembrane channels that permit facilitated diffusion can be opened or closed. They are said to be "gated".

Some types of gated ion channels:

- Ligand-gated
- Mechanically-gated
- Voltage-gated
- Light-gated

Ligand-gated ion channels: Many ion channels open or close in response to binding a small signaling molecule or "ligand". Some ion channels are gated by extracellular ligands; some by intracellular ligands. In both cases, the ligand is not the substance that is transported when the channel opens.

Examples:

- Acetylcholine (ACh). The binding of the neurotransmitter acetylcholine at certain synapses opens channels that admit Na+ and initiate a nerve impulse or muscle contraction.
- Gamma amino butyric acid (GABA). Binding of GABA at certain synapses — designated GABAA — in the central nervous system admits Cl- ions into the cell and inhibits the creation of a nerve impulse.
- Internal ligands

Internal ligands bind to a site on the channel protein exposed to the cytosol.

Examples:

- "Second messengers", like cyclic AMP (cAMP) and cyclic GMP (cGMP), regulate channels involved in the initiation of impulses in neurons responding to odors and light respectively.
- ATP is needed to open the channel that allows chloride (Cl-) and bicarbonate (HCO3-) ions out of the cell. This channel is defective in patients with cystic fibrosis. Although the energy liberated by the hydrolysis of ATP is needed to open the channel, this is not an example of active transport; the ions diffuse through the open channel following their concentration gradient.

Mechanically-gated-ion Channels: In so-called "excitable" cells like neurons and muscle cells, some channels open or close in response to changes in the charge (measured in volts) across the plasma membrane.

- Sound waves bending the cilia-like projections on the hair cells of the inner ear open up ion channels leading to the creation of nerve impulses that the brain interprets as sound.
- Mechanical deformation of the cells of stretch receptors opens ion channels leading to the creation of nerve impulses.
- Voltage-gated ion channels

Example: As an impulse passes down a neuron, the reduction in the voltage opens sodium channels in the adjacent portion of the membrane. This allows the influx of Na+ into the neuron and thus the continuation of the nerve impulse.

Some 7000 sodium ions pass through each channel during the brief period (about 1 millisecond) that it remains open. This was learned by use of the patch clamp technique.

The Patch Clamp Technique: The properties of ion channels can be studied by means of the patch clamp technique.

- A very fine pipette (with an opening of about 0.5 µm) is pressed against the plasma membrane of
- either an intact cell or
- the plasma membrane can be pulled away from the cell and the preparation placed in a test solution of desired composition.
- Current flow through a single ion channel can then be measured.

Such measurements reveal that each channel is either fully open or fully closed; that is, facilitated diffusion through a single channel is "all-or-none".

This technique has provided so much valuable information about ion channels that its inventors, Erwin Neher and Bert Sakmann, were awarded a Nobel Prize in 1991.

Facilitated Diffusion of Molecules

Some small, hydrophilic organic molecules, like sugars, can pass through cell membranes by facilitated diffusion.

Once again, the process requires transmembrane proteins. In some cases, these — like ion channels — form water-filled pores that enable the molecule to pass in (or out) of the membrane following its concentration gradient.

Example: Maltoporin. This homotrimer in the outer membrane of E. coli forms pores that allow the disaccharide maltose and a few related molecules to diffuse into the cell.

Another example: The plasma membrane of human red blood cells contain transmembrane proteins that permit the diffusion of glucose from the blood into the cell.

Note that in all cases of facilitated diffusion through channels, the channels are selective; that is, the structure of the protein admits only certain types of molecules through.

Whether all cases of facilitated diffusion of small molecules use channels is yet to be proven. Perhaps some molecules are passed through the membrane by a conformational change in the shape of the transmembrane protein when it binds the molecule to be transported.

Active Transport

Active transport is the pumping of molecules or ions through a membrane against their concentration gradient. It requires:

- A transmembrane protein (usually a complex of them) called a transporter and
- Energy: The source of this energy is ATP.

The energy of ATP may be used directly or indirectly.

- Direct Active Transport.: Some transporters bind ATP directly and use the energy of its hydrolysis to drive active transport.
- Indirect Active Transport: Other transporters use the energy already stored in the gradient of a directly-pumped ion. Direct active transport of the ion establishes a concentration gradient. When this is relieved by facilitated diffusion, the energy released can be harnessed to the pumping of some other ion or molecule.

Direct Active Transport

The Na+/K+ ATPase

The cytosol of animal cells contains a concentration of potassium ions (K+) as much as 20 times higher than that in the extracellular fluid. Conversely, the extracellular fluid contains a concentration of sodium ions (Na+) as much as 10 times greater than that within the cell.

These concentration gradients are established by the active transport of both ions. And, in fact, the same transporter, called the Na+/K+ ATPase, does both jobs. It uses the energy from the hydrolysis of ATP to

- Actively transport 3 Na+ ions out of the cell
- For each 2 K+ ions pumped into the cell.

This accomplishes several vital functions:

- It helps establish a net charge across the plasma membrane with the interior of the cell being negatively charged with respect to the exterior. This resting potential prepares nerve and muscle cells for the propagation of action potentials leading to nerve impulses and muscle contraction.
- The accumulation of sodium ions outside of the cell draws water out of the cell and thus enables it to maintain osmotic balance (otherwise it would swell and burst from the inward diffusion of water).
- The gradient of sodium ions is harnessed to provide the energy to run several types of indirect pumps.

The crucial roles of the Na+/K+ ATPase are reflected in the fact that almost one-third of all the energy generated by the mitochondria in animal cells is used just to run this pump.

The H+/K+ ATPase

The parietal cells of your stomach use this pump to secrete gastric juice. These cells transport protons (H+) from a concentration of about 4 x 10-8 M within the cell to a concentration of about 0.15 M in the gastric juice (giving it a pH close to 1). Small wonder that parietal cells are stuffed with mitochondria and uses huge amounts of energy as they carry out this three-million fold concentration of protons.

The Ca2+ ATPases

In resting skeletal muscle, there is a much higher concentration of calcium ions (Ca2+) in the sarcoplasmic reticulum than in the cytosol. Activation of the muscle fiber allows some of this Ca2+ to pass by facilitated diffusion into the cytosol where it triggers contraction.

After contraction, this Ca2+ is pumped back into the sarcoplasmic reticulum. This is done by a Ca2+ ATPase that uses the energy from each molecule of ATP to pump 2 Ca2+ ions.

A Ca2+ ATPase is also located in the plasma membrane of all eukaryotic cells. It pumps Ca2+ out of the cell helping to maintain the ~10,000-fold concentration gradient of Ca2+ between the cytosol (~ 10-7M) and the ECF (~ 10-3M).

Pumps 1. - 3. are designated P-type ion transporters because they use the same basic mechanism: a conformational change in the proteins as they are reversibly phosphorylated by ATP. And all three pumps can be made to run backward. That is, if the pumped ions are allowed to diffuse back through the membrane complex, ATP can be synthesized from ADP and inorganic phosphate.

ABC Transporters

ABC ("ATP-Binding Cassette") transporters are transmembrane protein that

- Expose a ligand-binding domain at one surface and a
- ATP-binding domain at the other surface.

The ligand-binding domain is usually restricted to a single type of molecule.

The ATP bound to its domain provides the energy to pump the ligand across the membrane.

The human genome contains 48 genes for ABC transporters. Some examples:

- CFTR – the cystic fibrosis transmembrane conductance regulator
- TAP, the transporter associated with antigen processing.

- The transporter that liver cells use to pump the salts of bile acids out into the bile.
- ABC transporters that pump chemotherapeutic drugs out of cancer cells thus reducing their effectiveness.

ABC transporters must have evolved early in the history of life. The ATP-binding domains in archaea, eubacteria, and eukaryotes all share a homologous structure, the ATP-binding "cassette".

Indirect Active Transport

Indirect active transport uses the downhill flow of an ion to pump some other molecule or ion against its gradient. The driving ion is usually sodium (Na+) with its gradient established by the Na+/K+ ATPase.

Symport Pumps

In this type of indirect active transport, the driving ion (Na+) and the pumped molecule pass through the membrane pump in the same direction.

Examples:

- The Na+/glucose transporter. This transmembrane protein allows sodium ions and glucose to enter the cell together. The sodium ions flow down their concentration gradient while the glucose molecules are pumped up theirs. Later the sodium is pumped back out of the cell by the Na+/K+ ATPase.
- The Na+/glucose transporter is used to actively transport glucose out of the intestine and also out of the kidney tubules and back into the blood.
- All the amino acids can be actively transported, for example
  - out of the kidney tubules and into the blood
  - the reuptake of Glu from the synapse back into the presynaptic neuron by sodium-driven symport pumps.

- The Na+/iodide transporter. This symporter pumps iodide ions into the cells of the thyroid gland (for the manufacture of thyroxine) and also into the cells of the mammary gland (to supply the baby's need for iodide).
- The permease encoded by the lac operon of E. coli that transports lactose into the cell.

Antiport Pumps

In antiport pumps, the driving ion (again, usually sodium) diffuses through the pump in one direction providing the energy for the active transport of some other molecule or ion in the opposite direction.

Example: Ca2+ ions are pumped out of cells by a sodium-driven antiport pump.

Antiport pumps in the vacuole of some plants harness the outward facilitated diffusion of protons (themselves pumped into the vacuole by a H+ ATPase)

- To the active inward transport of sodium ions. This sodium/proton antiport pump enables the plant to sequester sodium ions in its vacuole. Transgenic tomato plants that overexpress this sodium/proton antiport pump are able to thrive in saline soils too salty for conventional tomatoes.
- To the active inward transport of nitrate ions (NO_3).

Some Inherited Ion-channel Diseases

A growing number of human diseases have been discovered to be caused by inherited mutations in genes encoding channels.

Some examples:

- Chloride-channel diseases
  - Cystic fibrosis
  - Inherited tendency to kidney stones (caused by a different kind of chloride channel than the one involved in cystic fibrosis)

- Potassium-channel diseases
  - Some inherited life-threatening defects in the heartbeat
  - A rare, inherited tendency to epileptic seizures in the newborn.
  - Several types of inherited deafness
- Sodium-channel diseases
  - Inherited tendency to certain types of muscle spasms
  - Liddle's syndrome. Inadequate sodium transport out of the kidneys, because of a mutant sodium channel, leads to elevated osmotic pressure of the blood and resulting hypertension (high blood pressure).

Osmosis

- Osmosis is a special term used for the diffusion of water through cell membranes.
- Although water is a polar molecule, it is able to pass through the lipid bilayer of the plasma membrane. Transmembrane proteins that form hydrophilic channels accelerate the process, but even without these, water is still able to get through.
- Water passes by diffusion from a region of higher to a region of lower concentration. Note that this refers to the concentration of water, NOT the concentration of any solutes present in the water.
- Water is never transported actively; that is, it never moves against its concentration gradient. However, the concentration of water can be altered by the active transport of solutes and in this way the movement of water in and out of the cell can be controlled.

Example: the reabsorption of water from the kidney tubules back into the blood depends on the water following behind the active transport of Na+.

Hypotonic Solutions: If the concentration of water in the medium surrounding a cell is greater than that of the cytosol, the medium is said to be hypotonic. Water enters the cell by osmosis.

A red blood cell placed in a hypotonic solution (e.g., pure water) bursts immediately ("hemolysis") from the influx of water.

Plant cells and bacterial cells avoid bursting in hypotonic surroundings by their strong cell walls. These allow the buildup of turgor within the cell. When the turgor pressure equals the osmotic pressure, osmosis ceases.

Isotonic solutions: When red blood cells are placed in a 0.9% salt solution, they neither gain nor lose water by osmosis. Such a solution is said to be isotonic.

The extracellular fluid (ECF) of mammalian cells is isotonic to their cytoplasm. This balance must be actively maintained because of the large number of organic molecules dissolved in the cytosol but not present in the ECF. These organic molecules exert an osmotic effect that, if not compensated for, would cause the cell to take in so much water that it would swell and might even burst. This fate is avoided by pumping sodium ions out of the cell with the Na+/K+ ATPase.

Hypertonic Solutions: If red cells are placed in sea water (about 3% salt), they lose water by osmosis and the cells shrivel up. Sea water is hypertonic to their cytosol.

Similarly, if a plant tissue is placed in sea water, the cell contents shrink away from the rigid cell wall. This is called plasmolysis.

Sea water is also hypertonic to the ECF of most marine vertebrates. To avoid fatal dehydration, these animals (e.g., bony fishes like the cod) must

- Continuously drink sea water and then
- Desalt it by pumping ions out of their gills by active transport. (Marine reptiles – turtles and snakes – use special salt glands for the same purpose.)

A report of The New England Journal of Medicine tells of the life-threatening complications that can be caused by an ignorance of osmosis.

- Large volumes of a solution of 5% human albumin are injected into people undergoing a procedure called plasmapheresis.
- The albumin is dissolved in physiological saline (0.9% NaCl) and is therefore isotonic to human plasma (the large protein molecules of albumin have only a small osmotic effect).
- If 5% solutions are unavailable, pharmacists may substitute a proper dilution of a 25% albumin solution. Mixing 1 part of the 25% solution with 4 parts of diluent results in the correct 5% solution of albumin.
- BUT, in several cases, the diluent used was sterile water, not physiological saline.
- SO, the resulting solution was strongly hypotonic to human plasma.
- The Result: massive, life-threatening hemolysis in the patients.

CELL REPRODUCTION

Cell reproduction is asexual. The process of cell reproduction has three major parts. The first part of cell reproduction involves the replication of the parental cell's DNA. The second major issue is the separation of the duplicated DNA into two equally sized groups of chromosomess. The third major aspect of cell reproduction is the physical division of entire cells, usually called cytokinesis.

Cell reproduction is more complex in eukaryotes than in other organisms. Prokaryotic cells such as bacterial cells reproduce by binary fission, a process that includes DNA replication, chromosome segregation, and cytokinesis. Eukaryotic cell reproduction either involves mitosis or a more complex process called meiosis. Mitosis and meiosis are sometimes called the two "nuclear division" processes. Binary fission is similar to eukaryotic cell reproduction that involves mitosis. Both lead to the production of two daughter cells with the same number of chromosomes as the parental cell. Meiosis is used for a special cell reproduction process of diploid organisms. It produces four special daughter cells (gametes) which have half the normal cellular amount of DNA. A male and a female gamete can then combine to produce a zygote, a cell which again has the normal amount of chromosomes.

Comparison of the Three Types of Cell Reproduction

The DNA content of a cell is duplicated at the start of the cell reproduction process. Prior to DNA replication, the DNA content of a cell can be represented as the amount Z (the cell has Z chromosomes). After the DNA replication process, the amount of DNA in the cell is 2Z (multiplication: 2 x Z = 2Z). During Binary fission and mitosis the duplicated DNA content of the reproducing parental cell is separated into two equal halves that are destined to end up in the two daughter cells. The final part of the cell reproduction process is cell division, when daughter cells physically split apart from a parental cell. During meiosis, there are two cell division steps that together produce the four daughter cells.

After the completion of binary fission or cell reproduction involving mitosis, each daughter cell has the same amount of DNA (Z) as what the parental cell had before it replicated its DNA. These two types of cell reproduction produced two daughter cells that have the same number of chromosomes as the parental cell. After meiotic cell reproduction the four daughter cells have half the number of chromosomes that the parental cell originally had. This is the haploid amount of DNA, often symbolized as N. Meiosis is used by diploid organisms to produce haploid gametes. In a diploid organism such as the human organism, most cells of the body have the haploid amount of DNA, 2N. Using this notation for counting chromosomes we say that human somatic cells have 46 chromosomes (2N = 46) while human sperm and eggs have 23 chromosomes (N = 23). Humans have 23 distinct types of chromosomes, the 22 autosomes and the special category of sex chromosomes. There are two distinct sex chromosomes, the X chromosome and the Y chromosome. A diploid human cell has 23 chromosomes from that person's father and 23 from the mother. That is, your body has two copies of human chromosome number 2, one from each of your parents.

Chromosomes

Immediately after DNA replication a human cell will have 46 "double chromosomes". In each double chromosome there are two copies of that chromosome's DNA molecule. During mitosis the double chromosomes are split to produce 92 "single chromosomes", half of which go into each daughter cell. During meiosis, there are two chromosome separation steps which assure that each of the four daughter cells gets one copy of each of the 23 types of chromosome.

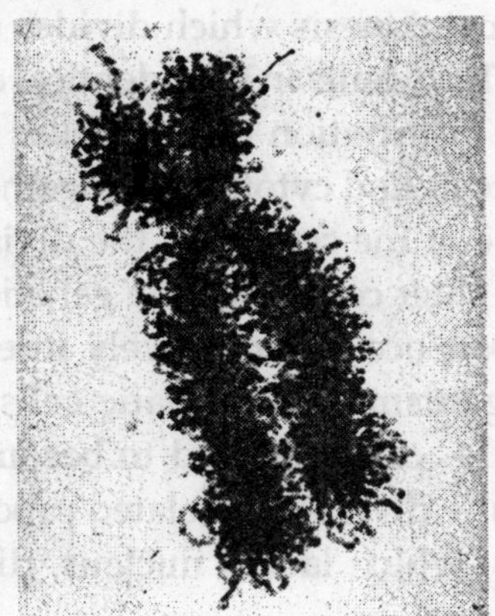

Fig. Chromosome

Sexual Reproduction

Though cell reproduction that uses mitosis can reproduce eukaryotic cells, eukaryotes bother with the more complicated process of meiosis because sexual reproduction such as meiosis confers a selective advantage. Notice that when meiosis starts, the two copies of chromosome number 2 are adjacent to each other. During this time, there can be genetic recombination events. Parts of the chromosome 2 DNA gained from one parent (red) will swap over to the chromosome 2 DNA molecule that received from the other parent (green). Notice that in mitosis the two copies of chromosome number 2 do not interact. It is these new combinations of parts of chromosomes that provide the major advantage for sexually reproducing organisms by allowing for new combinations of genes and more efficient evolution. However, in organisms with more than one set of chromosomes at the main life cycle stage, sex may also provide an advantage because, under random mating, it produces homozygotes and heterozygotes according to the Hardy-Weinberg ratio.

MITOSIS

Mitosis divides genetic information during cell division. Mitosis is the process by which a cell separates its duplicated genome into two identical halves. It is generally followed immediately by cytokinesis which divides the cytoplasm and cell membrane. This results in two identical daughter cells with a roughly equal distribution of organelles and other cellular components. Mitosis and cytokinesis together is defined as the mitotic (M) phase of the cell cycle, the division of the mother cell into two daughter cells, each the genetic equivalent of the parent cell. Mitosis occurs exclusively in eukaryotic cells. In multicellular organisms, the somatic cells undergo mitosis, while germ cells – cells destined to become sperm in males or ova in females – divide by a related process called meiosis. Prokaryotic cells, which lack a nucleus, divide by a process called binary fission.

The process of mitosis is complex and highly regulated. The sequence of events is divided into phases, corresponding to the completion of one set of activities and the start of the next. These stages are prophase, prometaphase, metaphase, telophase and anaphase. During the process of mitosis the pairs of chromosomes condense and attach to fibers that pull the sister chromatids to opposite sides of the cell. The cell then divides in cytokinesis, to produce two identical daughter cells.

Because cytokinesis usually occurs in conjunction with mitosis, "mitosis" is often used interchangeably with "mitotic phase". However, there are many cells where mitosis and cytokinesis occur separately, forming single cells with multiple nuclei. This occurs most notably among the fungi and slime moulds, but is found in various different groups. Even in animals, cytokinesis and mitosis may occur independently, for instance during certain stages of fruit fly embryonic development. Errors in mitosis can either kill a cell through apoptosis or cause mutations that may lead to cancer.

The primary result of mitosis is the division of the parent cell's genome into two daughter cells. The genome is composed of a number of chromosomes, complexes of tightly-coiled DNA that contain genetic information vital for proper cell function. Because each resultant daughter cell should be genetically identical to the parent cell, the parent cell must make a copy of each chromosome before mitosis. This occurs during the middle of interphase, the period that precedes the mitotic phase in the cell cycle where preparation for mitosis occurs.

Each chromosome now contains two identical copies of itself, called sister chromatids, attached together in a specialized region of the chromosome known as the centromere. Each sister chromatid is not considered a chromosome in itself, and a chromosome does not always contain two sister chromatids.

In most eukaryotes, the nuclear envelope that separates the DNA from the cytoplasm degrades, and its fluid spills out into the cytoplasm. The chromosomes align themselves in a line spanning the cell. Microtubules, essentially miniature strings, splay out from opposite ends of the cell and shorten, pulling apart the sister chromatids of each chromosome. As a matter of convention, each sister chromatid is now considered a chromosome, so they are renamed to sister chromosomes. As the cell elongates, corresponding sister chromosomes are pulled toward opposite ends. A new nuclear envelope forms around the separated sister chromosomes.

As mitosis completes cytokinesis is well underway. In animal cells, the cell pinches inward where the imaginary line used to be, separating the two developing nuclei. In plant cells, the daughter cells will construct a new dividing cell wall between each other. Eventually, the mother cell will be split in half, giving rise to two daughter cells, each with an equivalent and complete copy of the original genome.

Prokaryotic cells undergo a process similar to mitosis called binary fission. However, prokaryotes cannot be properly said to undergo mitosis because they lack a nucleus and only have a single chromosome with no centromere.

Phases

Interphase

The cell cycle: The mitotic phase is a relatively short action-packed period of the cell cycle. It alternates with the much longer interphase, where the cell prepares itself for division. Interphase is divided into three phases, G_1 (first gap), S (synthesis), and G_2 (second gap). During all three phases, the cell grows by producing proteins and cytoplasmic organelles. However, chromosomes are replicated only during the S phase. Thus, a cell grows (G_1), continues to grow as it duplicates its chromosomes (S), grows more and prepares for mitosis (G_2), and divides (M).

Mitosis is a continual and dynamic process. For purposes of description, however, mitosis is conventionally broken down into five subphases: prophase, prometaphase, metaphase, anaphase, and telophase, even though it is impossible to discern an exact "start" and "stop" of the phases.

Prophase

Prophase: The two round objects above the nucleus are the centrosomes. Note the condensed chromatin.

Normally, the genetic material in the nucleus is in a loosely bundled coil called chromatin. At the onset of prophase, chromatin condenses together into a highly ordered structure called a chromosome. Since the genetic material has already been duplicated earlier in S phase, the replicated chromosomes have two sister chromatids, bound together at the centromere by the protein cohesin. Chromosomes are visible at high magnification through a light microscope.

Just outside the nucleus are two centrosomes. Each centrosome, which was replicated earlier independent of mitosis, acts as a coordinating centre for the cell's microtubules. The two centrosomes sprout microtubules (which may be thought of as cellular ropes or poles) by polymerizing free-floating tubulin protein. By repulsive interaction of these microtubules with each other, the centrosomes push themselves to opposite ends of the cell (although new research has shown that there might be a mechanism inside the centromeres that also grab the microtubules and pull the chromatids apart). The network of microtubules is the beginning of the mitotic spindle.

Some centrosomes contain a pair of centrioles that may help organize microtubule assembly, but they are not essential to formation of the mitotic spindle.

In plant cells, prophase is preceded by a preprophase stage only found in plants. In highly vacuolated plant cells, the nucleus has to migrate into the centre of the cell before mitosis can begin. This is achieved through the formation of a phragmosome, a transverse sheet of cytoplasm that bisects the cell along the future plane of cell division. In addition to phragmosome formation, preprophase is characterized by the formation of a ring of microtubules and actin filaments (called preprophase band) underneath the plasmamembrane around the equatorial plane of the future mitotic spindle and predicting the position of cell plate fusion during telophase. The cells of higher plants (such as the flowering plants) lack centrioles. Instead, spindle microtubules aggregate on the surface of the nuclear envelope during prophase. The preprophase band disappears during nuclear envelope disassembly and spindle formation in prometaphase.

Prometaphase

The nuclear membrane has degraded, and microtubules have invaded the nuclear space. These microtubules can attach to kinetochores or they can interact with opposing microtubules.

The nuclear envelope dissolves and microtubules invade the nuclear space. This is called open mitosis, and it occurs in most multicellular organisms. Fungi and some protists, such as algae or trichomonads, undergo a variation called closed mitosis where the spindle forms inside the nucleus or its microtubules are able to penetrate an intact nuclear envelope.

Each chromosome forms two kinetochores at the centromere, one attached at each chromatid. A kinetochore is a complex protein structure that is analogous to a ring for the microtubule hook; it is the point where microtubules attach themselves to the chromosome. Although the kinetochore is not fully understood, it is known that it contains some form of molecular motor. When a microtubule connects with the kinetochore, the motor activates, using energy from ATP to "crawl" up the tube toward the originating centrosome. This motor activity, coupled with polymerisation and depolymerisation of microtubules, provides the pulling force necessary to later separate the chromosome's two chromatids. Prometaphase is sometimes considered part of prophase.

Metaphase

The chromosomes have aligned at the metaphase plate. As microtubules find and attach to kinetochores in prometaphase, the centromeres of the chromosomes convene along the metaphase plate or equatorial plane, an imaginary line that is equidistant from the two centrosome poles. This even alignment is due to the counterbalance of the pulling powers generated by the opposing kinetochores, analogous to a tug-of-war between equally strong people. In certain types of cells, chromosomes do not line up at the metaphase plate and instead move back and forth between the poles randomly, only roughly lining up along the midline.

Because proper chromosome separation requires that every kinetochore be attached to a bundle of microtubules (spindle fibers), it is thought that unattached kinetochores generate a signal to prevent premature progression to anaphase without all chromosomes being aligned. The signal creates the mitotic spindle checkpoint.

Anaphase

Early anaphase: Kinetochore microtubules shorten. When every kinetochore is attached to a cluster of microtubules and the chromosomes have lined up along the metaphase plate, the cell proceeds to anaphase.

Two events then occur; First, the proteins that bind sister chromatids together are cleaved, allowing them to separate. These sister chromatids turned sister chromosomes are pulled apart by shortening kinetochore microtubules and toward the respective centrosomes to which they are attached. Next, the nonkinetochore microtubules elongate, pushing the centrosomes (and the set of chromosomes to which they are attached) apart to opposite ends of the cell.

These two stages are sometimes called early and late anaphase. At the end of anaphase, the cell has succeeded in separating identical copies of the genetic material into two distinct populations.

Telophase

The pinching is known as the cleavage furrow. Telophase is a reversal of prophase and prometaphase events. It "cleans up" the aftereffects of mitosis. At telophase, the nonkinetochore microtubules continue to lengthen, elongating the cell even more. Corresponding sister chromosomes attach at opposite ends of the cell. A new nuclear envelope, using fragments of the parent cell's nuclear membrane, forms around each set of separated sister chromosomes. Both sets of chromosomes, now surrounded by new nuclei, unfold back into chromatin.

Cytokinesis

Often (mistakenly) thought to be the same process as telophase, cytokinesis, if it is to occur, is usually well under way by this time. In animal cells, a cleavage furrow (pinch) containing a contractile ring develops where the metaphase plate used to be, pinching off the separated nuclei. In both animal and plant cells, cell division is also driven by vesicles derived from the Golgi apparatus, which move along microtubules to the middle of the cell. In plants this structure coalesces into a cell plate at the centre of the phragmoplast and develops into a cell wall, separating the two nuclei. The phragmoplast is a microtubule structure typical for higher plants, whereas some green algae use a phycoplast microtubule array during cytokinesis. Each daughter cell has a complete copy of the genome of its parent cell. Mitosis is complete.

Summary: The final step of cell division in which the membrane cleaves and there are two separate cells. The original cell has stopped dividing.

Significance: The importance of mitosis is the maintenance of the chromosomal set. Transcription is generally believed to cease during mitosis, but epigenetic mechanisms such as bookmarking function during this stage of the cell cycle to ensure that the "memory" of which genes were active prior to entry into mitosis are transmitted to the daughter cells.

MEIOSIS AND SEXUAL REPRODUCTION

MEIOSIS

Sexual reproduction occurs only in eukaryotes. During the formation of gametes, the number of chromosomes is reduced by half, and returned to the full amount when the two gametes fuse during fertilization.

Ploidy

Haploid and diploid are terms referring to the number of sets of chromosomes in a cell. Gregor Mendel determined his peas had two sets of alleles, one from each parent. Diploid organisms are those with two (di) sets. Human beings (except for their gametes), most animals and many plants are diploid. Here abbreviate diploid as 2n. Ploidy is a term referring to the number of sets of chromosomes. Haploid organisms/cells have only one set of chromosomes, abbreviated as n. Organisms with more than two sets of chromosomes are termed polyploid. Chromosomes that carry the same genes are termed homologous chromosomes. The alleleds on homologous chromosomes may differ, as in the case of heterozygous individuals. Organisms (normally) receive one set of homologous chromosomes from each parent.

Meiosis is a special type of nuclear division which segregates one copy of each homologous chromosome into each new "gamete". Mitosis maintains the cell's original ploidy level (for example, one diploid 2n cell producing two diploid 2n cells; one haploid n cell producing two haploid n cells; etc.). Meiosis, on the other hand, reduces the number of sets of chromosomes by half, so that when gametic recombination (fertilization) occurs the ploidy of the parents will be reestablished.

Most cells in the human body are produced by mitosis. These are the somatic (or vegetative) line cells. Cells that become gametes are referred to as germ line cells. The vast majority of cell divisions in the human body are mitotic, with meiosis being restricted to the gonads.

Life Cycles

Life cycles are a diagrammatic representation of the events in the organism's development and reproduction. When interpreting life cycles, pay close attention to the ploidy level of particular parts of the cycle and where in the life cycle meiosis occurs. For example, animal life cycles have a dominant diploid phase, with the gametic (haploid) phase being a relative few cells. Most of the cells in your body are diploid, germ line diploid cells will undergo meiosis to produce gametes, with fertilization closely following meiosis.

Plant life cycles have two sequential phases that are termed alternation of generations. The sporophyte phase is "diploid", and is that part of the life cycle in which meiosis occurs. However, many plant species are thought to arise by polyploidy and the use of "diploid" in the last sentence was meant to indicate that the greater number of chromosome sets occur in this phase. The gametophyte phase is "haploid", and is the part of the life cycle in which gametes are produced (by mitosis of haploid cells). In flowering plants (angiosperms) the multicelled visible plant (leaf, stem, etc.) is sporophyte, while pollen and ovaries contain the male and female gametophytes, respectively. Plant life cycles differ from animal ones by adding a phase (the haploid gametophyte) after meiosis and before the production of gametes.

Many protists and fungi have a haploid dominated life cycle. The dominant phase is haploid, while the diploid phase is only a few cells (often only the single celled zygote, as in Chlamydomonas). Many protists reproduce by mitosis until their environment deteriorates, then they undergo sexual reproduction to produce a resting zygotic cyst.

Phases of Meiosis

Two successive nuclear divisions occur, Meiosis I (Reduction) and Meiosis II (Division). Meiosis produces 4 haploid cells. Mitosis produces 2 diploid cells. The old name for meiosis was reduction/division. Meiosis I reduces the ploidy level from 2n to n (reduction) while Meiosis II divides the remaining set of chromosomes in a mitosis-like process (division). Most of the differences between the processes occur during Meiosis I.

Prophase I

Prophase I has a unique event — the pairing (by an as yet undiscovered mechanism) of homologous chromosomes. Synapsis is the process of linking of the replicated homologous chromosomes. The resulting chromosome is termed a tetrad, being composed of two chromatids from each chromosome, forming a thick (4-strand) structure. Crossing-over may occur at this point. During crossing-over chromatids break and may be reattached to a different homologous chromosome.

The alleles on this tetrad:

A B C D E F G
A B C D E F G

a b c d e f g

a b c d e f g

will produce the following chromosomes if there is a crossing-over event between the 2nd and 3rd chromosomes from the top:

A B C D E F G

A B c d e f g

a b C D E F G

a b c d e f g

Thus, instead of producing only two types of chromosome (all capital or all lower case), four different chromosomes are produced. This doubles the variability of gamete genotypes. The occurrence of a crossing-over is indicated by a special structure, a chiasma (plural chiasmata) since the recombined inner alleles will align more with others of the same type (e.g. a with a, B with B). Near the end of Prophase I, the homologous chromosomes begin to separate slightly, although they remain attached at chiasmata.

Crossing-over between homologous chromosomes produces chromosomes with new associations of genes and alleles.

Events of Prophase I (save for synapsis and crossing over) are similar to those in Prophase of mitosis: chromatin condenses into chromosomes, the nucleolus dissolves, nuclear membrane is disassembled, and the spindle apparatus forms.

Metaphase I

Metaphase I is when tetrads line-up along the equator of the spindle. Spindle fibers attach to the centromere region of each homologous chromosome pair. Other metaphase events as in mitosis.

Anaphase I

Anaphase I is when the tetrads separate, and are drawn to opposite poles by the spindle fibers. The centromeres in Anaphase I remain intact.

Telophase I

Telophase I is similar to Telophase of mitosis, except that only one set of (replicated) chromosomes is in each "cell". Depending on species, new nuclear envelopes may or may not form. Some animal cells may have division of the centrioles during this phase.

Prophase II

During Prophase II, nuclear envelopes (if they formed during Telophase I) dissolve, and spindle fibers reform. All else is as in Prophase of mitosis. Indeed Meiosis II is very similar to mitosis.

Metaphase II

Metaphase II is similar to mitosis, with spindles moving chromosomes into equatorial area and attaching to the opposite sides of the centromeres in the kinetochore region.

Anaphase II

During Anaphase II, the centromeres split and the former chromatids (now chromosomes) are segregated into opposite sides of the cell.

Telophase II

Telophase II is identical to Telophase of mitosis. Cytokinesis separates the cells.

Comparison of Mitosis and Meiosis

Mitosis maintains ploidy level, while meiosis reduces it. Meiosis may be considered a reduction phase followed by a slightly altered mitosis. Meiosis occurs in a relative few cells of a multicellular organism, while mitosis is more common.

Gametogenesis

Gametogenesis is the process of forming gametes (by definition haploid, n) from diploid cells of the germ line. Spermatogenesis is the process of forming sperm cells by meiosis (in animals, by mitosis in plants) in specialized organs known as gonads (in males these are termed testes). After division the cells undergo differentiation to become sperm cells. Oogenesis is the process of forming an ovum (egg) by meiosis (in animals, by mitosis in the gametophyte in plants) in specialized gonads known as ovaries. Whereas in spermatogenesis all 4 meiotic products develop into gametes, oogenesis places most of the cytoplasm into the large egg. The other cells, the polar bodies, do not develop. This all the cytoplasm and organelles are go into the egg. Human males produce 200,000,000 sperm per day, while the female produces one egg (usually) each menstrual cycle.

Spermatogenesis

Sperm production begins at puberty at continues throughout life, with several hundred million sperm being produced each day. Once sperm form they move into the epididymis, where they mature and are stored.

Oogenesis

The ovary contains many follicles composed of a developing egg surrounded by an outer layer of follicle cells. Each egg begins oogenesis as a primary oocyte. At birth each female carries a lifetime supply of developing oocytes, each of which is in Prophase I. A developing egg (secondary oocyte) is released each month from puberty until menopause, a total of 400-500 eggs.

Chapter 6

The Chemistry of Biology

Much of biology is based on the principles of chemistry; as was mentioned in earlier chapter, cells are composed of organic molecules, so a discussion of cells would not be complete without talking about these substances. Chemistry is also important in that organisms perform chemical reactions to store and release the energy that they use to carry out their various processes, and the way cell organelles work has a simple basis in chemistry. Chapter focuses on the fundamental structure of matter, the manner in which chemical reactions occur, the different types of molecules as well as their classification, and the ways in which different molecules can interact with one another. If you have a background in chemistry, feel free to either skip this chapter or skim through the section titles to make sure that you're familiar with the material.

ELEMENTS AND ATOMS (MH)

Everything, whether it be a giraffe, a plastic cup, your computer, or the air, is made of what is called matter. Matter is just the "stuff" that everything, absolutely everything, is made of. Certain types of matter can be broken down into simpler pieces of matter, some of which can be broken down again and again. The process of breaking down (or building up) matter is performed by a chemical reaction; that is, when two substances (chemicals) come together and react with one another. Why this occurs will become clear to you later.

When matter is broken down into simpler and simpler parts, there comes a point when chemical reactions can no longer split the matter. This special type of matter that cannot be further broken is called an element. For example, oxygen, hydrogen, carbon, and sodium are the names of some commonly known elements. Each element has a one or two letter symbol which is used by scientists to identify it. Using the previous examples, oxygen is O, hydrogen is H, carbon is C, and sodium is Na.

Currently, over one hundred elements have been discovered, and each has its own distinct properties. Any amount of a particular element, whether it be the size of a car or the size of a pea, will always exhibit those same properties. The smallest amount of an element that you can have that still exhibits these properties is called an atom, the unit of structure of matter.

Atoms are very, very small. Even the most powerful microscopes can only picture atoms as tiny white dots! Since observing atoms directly is nearly impossible, scientists have, over the years, determined the basic structure of an atom through experimentation. All atoms share a similar structure. In the centre is a nucleus, a ball of two types of particles called protons and neutrons. Both of these particles have the same mass (one atomic mass unit), but the proton has a positive charge while the neutron has a neutral charge. Similar to how the planets orbit the Sun, tiny particles called electrons orbit the nucleus. Electrons are much smaller than protons, but they happen to have a negative charge exactly equal to the positive charge of a proton. Normal atoms contain the same number of protons and electrons, so the positive and negative charges cancel, leaving the entire atom with neutral charge.

What makes atoms different from one another is the number of protons, neutrons, and electrons that they contain. All atoms of a particular element have the same number of each of these particles. Scientists have therefore given each element an atomic number, which is simply equal to the number of protons in an atom of that element. For example, all atoms of the simplest element, hydrogen, have one proton and one electron. The atomic number of hydrogen is 1. A sodium atom has eleven protons and twelve neutrons in the nucleus, which is orbited by eleven electrons. The atomic number of sodium is 11. The atomic mass of an element is just the number of protons plus the number of neutrons it has. So, the atomic mass of hydrogen is 1 (1 proton + 0 neutrons = 1), and the atomic mass of sodium is 23 (11 protons + 12 neutrons = 23).

An important point to remember is that positive and negative charges attract to one another, while two negative charges or two positive charges repel one another. Have you wondered why there are only about a hundred elements? Well, protons in the nucleus of an atom tend to repel one another (because they all have a positive charge), so the nucleus is slightly unstable. In elements with a great many protons, the force of repulsion is so great that the nucleus breaks apart. Although over a hundred elements have been discovered, many of them must be artificially created in a laboratory, and they quickly break apart once they are formed.

Certain elements come in several forms, but each of these forms differs from the first only in the number of neutrons in the nucleus. These separate forms are called isotopes of that element. For instance, hydrogen has two isotopes, deuterium and tritium. Normally, hydrogen has no neutrons, but deuterium has a single neutron in the nucleus, and tritium has two neutrons.

MOLECULES AND THEIR FORMATION

Through chemical reactions, atoms can combine and form complex structures called molecules. The atoms in molecules stay together due to forces called bonds between the atoms. Bonds come in two major types: ionic bonds and covalent bonds. A full understanding of these bonds is not necessary for understanding the biology principles that rest on them, but knowing the distinction between them is still important.

One important thing to know about bonds is that they have nothing to do with gravity. Gravity is a force of attraction between all atoms. Objects which are very massive (the Earth, for example) exert a very large gravitational force. That's why we stick to the ground. Objects with very little mass, like the protons, neutrons, and electrons in an atom, produce an incredibly tiny gravitational force. This miniscule force does not play any role in the formation of bonds between atoms.

Sometimes, atoms will become more stable if they either obtain or lose electrons. For example, the element chlorine becomes more stable when it gains an electron, and sodium becomes more stable when it loses an electron. If an atom of chlorine and an atom of sodium come near enough to one another, an electron from the sodium can actually jump to the chlorine atom. However, the sodium atom now has more protons than electrons, so it has a positive charge. The chlorine atom now has an additional electron and so has a negative charge. These charged atoms are called ions. Ions of opposite charges, like the sodium and chlorine ions, are attracted to one another in what is called an ionic bond.

In other cases, two atoms will share the electrons so that each becomes more stable. For example, oxygen, which normally has eight electrons, happens to become more stable when it has two additional electrons. If two oxygen atoms come together, they can each contribute a pair of electrons to be shared by both atoms. This type of sharing is called a covalent bond. In the case of the oxygen atoms, the electrons are shared exactly equally. However, in most cases, one atom pulls on the electrons more than the other. In these bonds, one side of the molecule is slightly more negative since the negatively charged electrons are closer to that side. These molecules are said to be polar (water is an example), whereas molecules which share the electrons equally are nonpolar. Polar and nonpolar molecules are very significant in how many aspects of unicellular organisms function.

MOLECULAR AND STRUCTURAL FORMULAS

As mentioned in the previous section, molecules are made up of different types of atoms which join with one another through some sort of bond, usually either ionic or covalent. Molecules are often made of many different types of atoms; for example, a chemical called glucose contains six carbon atoms, twelve hydrogen atoms, and six oxygen atoms. Scientists have adopted two ways of expressing what a particular molecule consists of: molecular formulas and structural formulas.

Molecular formulas are useful in that they are a very compact and simple way of writing complex molecules. As mentioned earlier, each element has a one or two letter symbol which is used to identify it. In a molecular formula, all of the symbols for the elements that are contained in the molecule are written, and to the right of each is a number which indicates how many atoms of that element are used. For example, glucose has six carbon atoms, twelve hydrogen atoms, and six oxygen atoms. The symbols for carbon, hydrogen, and oxygen are C, H, and O respectively, so the molecular formula for glucose is $C_6H_{12}O_6$. As another example, water has two hydrogen atoms and one oxygen atom, so its molecular formula is H_2O. Notice that there is no "1" next to the O in the molecular formula for water. Scientists have decided that if there is just one atom of that element, then the number "1" does not need to be written next to the element in the molecular formula.

Sometimes a scientist may want to say that he or she has not one but several molecules of the same type. The method of writing this is very simple. To indicate that you have 8 molecules of water, you just write $8H_2O$. Notice that there are 16 total hydrogen atoms here, since each molecule of water has 2 and there are 8 molecules. Counting atoms like this is sometimes important when writing chemical reactions (which are discussed in the next section).

The only problem with molecular formulas is that they do not say how the atoms in the molecule are arranged; that is, which atom is bonded to which other atom. That's why scientists also use structural formulas. Structural formulas are actual sketches of the molecule, which indicate where each atom is placed. In a structural formula, each atom is represented by its symbol, and bonds are indicated by lines connecting the atoms. Occasionally, two molecules will have the exact same number of their types of atoms, even though their structure is different. In these cases, the molecular formulas for the molecules are identical, so only the structural formula can show the difference between them.

CHEMICAL REACTIONS

A chemical reaction is simply the breaking of bonds and/ or the formation of new bonds between atoms and molecules. Chemical reactions are written in a very simple way; they look somewhat like a math equation. On the left side are the compounds (compounds are just collections of the same type of molecule) that existed before the reaction, and on the right side are the compounds which formed after the reaction. These compounds are called the reactants and the products respectively. Below is a simple chemical reaction that often occurs within cells.

$$C_6H_{12}O_6 + 6O_2 \longrightarrow 6CO_2 + 6H_2O + \text{energy}$$

Now, let's look at this reaction in further detail. The reactants in this reaction are $C_6H_{12}O_6$ (also known as glucose, a type of sugar) and O_2 (the form of oxygen one would find in the atmosphere). Notice that one molecule of glucose and six molecules of O_2 are required for the reaction. The products of the reaction are CO_2 (carbon dioxide) and H_2O (water), six molecules of each. Also produced is energy, which is released when some of the bonds in the glucose molecule are broken.

Try counting how many atoms of carbon, oxygen, and hydrogen there are on either side of the reaction. You'll find that there are 6 carbon atoms, 18 oxygen atoms, and 12 hydrogen atoms on both sides. In all chemical reactions in a cell, the number of atoms of each element on either side of the reaction is equal

ACIDS, BASES, AND BUFFERS

Acids and bases are a way of classifying compounds based upon what happens to them when you place them in water.

Acids are substances that, when dissolved in water, split into two ions, one of which is an H+ ion (The H indicates that the ion is hydrogen, and the + indicates that it is positively charged, meaning that there is no electron.). A well known acid is HCl (hydrochloric acid), which splits into two ions when placed in water: H^+ and Cl^-.

When a base is dissolved in water, it splits into ions as well. For a base, one of them is OH- (often called a hydroxide ion). For example, NaOH splits into Na+ and OH- when placed in water, so it is a base.

If you put both an acid and a base into the same container of water, they tend to cancel out the effects of one another. For example, if both HCl and NaOH are placed in water, the Na+ and Cl- ions combine to form NaCl (table salt), and the H+ and OH- ions combine to form H_2O (water).

Acids that release more H+ ions than other are said to be more acidic, and bases which release greater amounts of OH- ions are more basic. The extent to which a compound is acidic or basic is measured by the pH scale. On the scale, numbers range from 0 (most acidic) to 14 (most basic). Water has a pH of 7, which is neutral.

Buffers are compounds that tend to neutralize the pH of a solution by combining with either H+ ions or OH- ions to keep the solution neutral. Buffers play a very important role in most organisms, as many organisms cannot live at pHs that are too acidic or too basic. This is because certain reactions which occur in organisms are hindered by the effects of an excess of charged ions in the environment

ORGANIC COMPOUNDS AND THE IMPORTANCE OF CARBON

Organic molecules are any molecules that contain atoms from three elements: carbon, hydrogen, and oxygen. For example, glucose is organic, since its molecular formula is $C_6H_{12}O_6$. Carbon dioxide (CO_2) is inorganic since it does not contain hydrogen.

All organic molecules have two important parts, the carbon backbone and the functional groups. To understand why carbon is the backbone of most molecules, we have to think back to the topic of bonds between atoms. Carbon becomes more stable when it gains four more electrons; that is, when it makes four bonds. So, every carbon atom in an organic molecule has made four bonds, which allows the molecule to have a complex structure with carbon as the "backbone."

At each end of the molecule is a functional group. Often organic compounds are classified based upon their functional groups because molecules with the same functional groups have similar properties. In general, when organic molecules chemically combine to form larger molecules, the bonding occurs at the functional group. So, certain molecules combine frequently with one another because their functional groups are "compatible." The most important functional groups are shown below. Later, you'll understand how certain pairs of these groups can react with one another to join two molecules together.

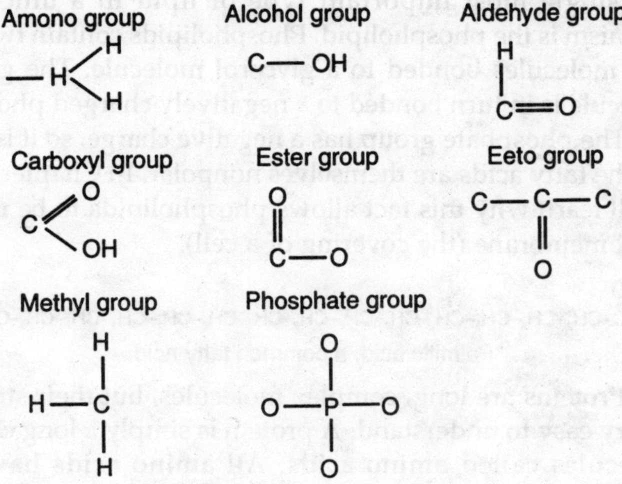

DIFFERENT TYPES OF ORGANIC COMPOUNDS

Mentioned in the previous section, organic molecules are classified based upon their functional groups. Here we will discuss three of the most important types of organic compounds: carbohydrates, lipids, and proteins.

Carbohydrates are molecules formed through the joining of sugar molecules. Sugars, such as glucose and fructose, always have the same number of carbon atoms and oxygen atoms, and have twice as many hydrogen atoms. The previous examples are known as monosaccharides (this is the simplest type of sugar); a sugar consisting of two monosaccharides is called a disaccharide. Monosaccharides and disaccharides can link together to form long chains which are known as polysaccharides. Carbohydrates are often used for both the storage of energy and the building blocks for the structure of many cell organelles.

Lipids, sometimes known as fats, are a class of organic molecules which do not dissolve in water and other polar compounds. Most lipids consist of molecules known as fatty acids. Lipids generally do not dissolve in water, so they are said to be hydrophobic which literally means "afraid of water." The single most important type of lipid in a unicellular organism is the phospholipid. Phospholipids contain two fatty acid molecules bonded to a glycerol molecule. The glycerol molecule is in turn bonded to a negatively charged phosphate ion. The phosphate group has a negative charge, so it is polar, but the fatty acids are themselves nonpolar. In Chapter Three, you'll learn why this fact allows phospholipids to be used as a cell membrane (the covering of a cell).

$$OH{-}\overset{\overset{\large O}{\|}}{C}{-}CH_2{-}CH_2{-}CH_2{-}CH_2{-}CH_2{-}CH_2{-}CH_2{-}CH_2{-}CH_2{-}CH_2{-}CH_2{-}CH_2{-}CH_2{-}CH_2{-}CH_3$$

Palmitic acid, a common fatty acid

Proteins are long, complex molecules, but their structure is very easy to understand. A protein is simply a long chain of molecules called amino acids. All amino acids have two functional groups, an amino group (NH_2) and a carboxyl group

(COOH), but each amino acid also has a group which is often called the variable group since is varies from amino acid to amino acid. Amino acids bond with each other through what is known as a peptide bond. The OH from the carboxyl group and the H from the amino group break off and join together to form OH + H = H_2O (water!). Meanwhile, the C in the carboxyl group and the N in the amino group from a bond (the peptide bond). This process, where water is removed and two molecules are joined together, is called dehydration synthesis.

ENZYMES AND CO-ENZYMES

Enzymes are very special types of proteins for all living things; they are called organic catalysts. The first part makes sense, since enzymes are proteins, and proteins are organic, but let's make clear what it means to be a catalyst. A catalyst is any substance that speeds up the rate of a chemical reaction but is itself not affected once the reaction is completed. Enzymes are used in organisms to increase the rate of the chemical reactions that are necessary for life.

For an enzyme to catalyze a reaction, it must join with one or more of the molecules in the reaction; the molecules that an enzyme attaches to are called substrates, and when they join they form an enzyme-substrate complex. However, enzymes are able to attach only to certain substrates, a fact which is explained by the structure of an enzyme. Enzymes are proteins which have folded onto themselves several times to create a complex, three-dimensional structure. Each enzyme has an area called the active site where the substrate will join, but, like a lock and a key, only certain substrates will fit into the active site.

Unfortunately, it's not as simple as just a lock and a key. Sometimes, a substrate doesn't fit exactly into the active site of an enzyme, but the match is fairly close. In these cases, the enzyme is induced (persuaded) to change the shape of its active site slightly so that the substrate fits. This explanation of an enzyme's activity is called the "induced-fit hypothesis" and is generally accepted by most biologists.

Coenzymes are organic molecules which are not proteins like enzymes but still play a role in reactions catalyzed by enzymes. In cells, coenzymes frequently serve as electron acceptors; they bond with electrons released by chemical reactions in the cell.

FACTORS WHICH AFFECT THE EFFICIENCY OF AN ENZYME

There are a number of influences than can change the efficiency of enzyme. We will discuss the four most important factors: inhibitors, allosteric factors, pH, and temperature.

There are two types of inhibitors: competitive inhibitors and noncompetitive inhibitors, and their names give a good indication of what they actually do. Competitive inhibitors have a similar structure to the enzyme's substrate, so they can "compete" with the substrate for the active site of an enzyme. Often the enzyme will bond not to its substrate but to the competitive inhibitor, blocking the substrate from the active site and causing the formation of enzyme-substrate complexes to occur at a slower rate. Noncompetitive inhibitors, on the other hand, do not attach to the active site and block the enzyme-substrate complex from forming. Instead, they react with portions of the active site, which results in the changing of its shape. Once the active site's shape is changed, it can no longer attach to the substrate.

Some enzymes have special areas other than the active site. These special areas are sometimes called regulatory sites. Any molecule that attaches to the regulatory site is called an allosteric factor. Allosteric inhibitors join with the regulatory site and change the shape of the entire enzyme (including the active site), thus preventing it from binding with the substrate. However, not all allosteric factors are detrimental. Some join with the regulatory site and actually bring the enzyme and its active site into the proper shape so that it can successfully form enzyme-substrate complexes.

pH also plays an important role in affecting the rate of an enzyme's activity. Remember that pH is a measure of how acidic or basic a solution is; that is, how many H+ or OH- ions there are. These ions are charged, and charged molecules tend to pull on other molecules. So, if too many ions are present, the enzyme may be denatured (twisted and pulled so out of shape that it can no longer function). However, this is not to say that all enzymes work best when the pH is neutral. Some enzymes actually work best in acidic or basic environments, but these characteristics are particular to the enzyme.

The final factor that influences an enzyme's efficiency is temperature. To a certain extent, a high temperature increases the rate of an enzyme's activity, because at high temperatures, molecules move around faster, so an enzyme is likely to come in contact with a substrate very quickly. However, at too high temperatures, the enzyme can become denatured and lose all function. Low temperatures slow the rate of formation of the enzyme-substrate complex because the molecules move at slower speeds and so do not come in contact with one another as frequently.

DIFFUSION

Diffusion is a phenomenon which is essential to all unicellular organisms. Most people have had firsthand experience with diffusion without knowing it, but it's easy to see it in action. Have you ever been in another room, yet you were still able to smell the food cooking in the kitchen? You may not know it, but that's because of diffusion. Here's another example you can try. Take a cup of coffee without any milk in it. Using a spoon, drop a little bit of milk by one edge of the cup and watch as the milk spreads through the coffee. This example is also caused by the process of diffusion.

The scientific definition of diffusion is "the movement of molecules from an area of high concentration to an area of low concentration." Since the food was cooking in the kitchen, most of the molecules which cause the smell are located there (that is, the kitchen has a high concentration of molecules), and very few of them are located in the next room. By diffusion, the molecules which cause smell tend to move out of the kitchen and into other rooms. Similarly, the milk was highly concentrated near the edge where you placed it, so it diffused to areas of lower concentration, namely the rest of the cup. The difference in concentration in both of the examples is sometimes referred to as a concentration gradient. Molecules move "down" a concentration gradient; that is, they move toward areas with fewer molecules.

Gradients do not necessarily have to be of concentration. For example, if one side of a membrane is highly acidic while the other side is neutral we say that there is a pH gradient. The acidic molecules will tend to move "down" a pH gradient; that is, they move toward the neutral side until both sides are equally acidic.

The diffusion of water has a special name: osmosis. Most unicellular organisms live in aquatic environments such as an ocean or lake. The water outside of the cell is usually fairly pure, whereas the water inside the cell contains all sorts of organic and inorganic molecules. By diffusion, these molecules will tend to move out of the cell since they are less concentrated in the external environment. Also, the water in the environment will tend to move into the cell by osmosis. To prevent themselves from bursting because of too much water, many cells have developed specialized structures to keep excess water out of the cell.

DNA, RNA, AND PROTEIN SYNTHESIS

DNA is arguably the most intriguing aspect of biology: a single molecule acting as a genetic code for the production of proteins in a cell. How DNA accomplishes this function is truly remarkable. This chapter begins with a brief introduction to DNA and the history of the discovery of its significance and then moves on to discuss the structure of DNA and the mechanisms by which it indirectly controls protein synthesis.

EARLY HYPOTHESES REGARDING THE GENETIC MATERIAL

Most people look somewhat like a mixture of their parents. They may have their father's nose or their mother's eyes, but in general, certain traits are passed on from one generation to the next. This is also the case with unicellular organisms. This "genetic information" (the information which determines the traits which are passed down) must be stored somewhere in the cell, but just how or where eluded biologists for a long time.

Eventually, microscopists noticed that during cell division, the sister chromatid pairs split, and each corresponding chromosome became part of a different daughter cell. This phenomenon suggested to biologists that the chromosomes must in some way contain the genetic information, especially since cells which obtained an abnormal number of chromosomes failed to function. However, chemical analysis showed that the chromosomes were made of two major components: DNA (deoxyribonucleic acid) and protein.

DNA was, at the time, thought to be a much simpler molecule than proteins, which scientists knew came in many varieties and combinations. So, it seemed logical that they hypothesized that protein was the genetic material. But by the late 1920s and 1930s, experimental evidence (discussed in the next section) began to mount that DNA was in fact the bearer of genetic information.

THE FAMOUS DNA EXPERIMENTS

In 1928, Fred Griffith performed the first experiment which suggested that protein was not the genetic material. His experiment was actually fairly simple. He first injected mice with a live strain of virulent (deadly) bacteria, and not to anyone's surprise, all of those mice died. Then, he killed the virulent bacteria cells by heating them. Mice injected with these heat-killed virulent bacteria did not die. In another set of mice, Griffith injected a live non-virulent strain of bacteria, and these mice did not die, the result which Griffith expected.

The surprise came when Griffith injected a group of mice with both live non-virulent bacteria and heat-killed virulent bacteria. In that group, some of the mice died. When Griffith examined those mice, he found live virulent bacteria in their blood. Griffith drew the conclusion that the genetic information in the heat-killed virulent bacteria survived the heating process and was somehow incorporated into the genetic material of the non-virulent strain to cause them to become virulent. But Griffith knew that heat denatures protein, so he suggested that the genetic material must be something else. However, his results did not specifically point to DNA as a possibility.

Oswald Avery followed up on Griffith's experiment in the following decade. Like Griffith, Avery first used heat to kill virulent bacteria. He then extracted RNA (ribonucleic acid), DNA, carbohydrates, lipids, and proteins from these dead cells, all of which were considered to be possible candidates for the carriers of genetic information. Next, he added each type of molecule to a culture of live non-virulent bacteria to determine which was responsible for changing them into virulent bacteria as Griffith had observed. Only the non-virulent cells which were given DNA from the dead virulent strain became virulent, so Avery concluded that DNA must be the genetic material.

DNA'S CHEMICAL COMPOSITION AND STRUCTURE

Chemical analyses by scientists revealed the general chemical composition of nucleic acids (DNA and RNA): they are composed of nucleotides. A nucleotide consists of a phosphate group, a five-carbon sugar (deoxyribose in DNA and ribose in RNA), and a nitrogenous base bonded together.

Each nucleotide in a DNA molecule has one of four nitrogenous bases: adenine, guanine, thymine, and cytosine. The first two are called purine bases because their structure consists of two rings of atoms. The latter two are known as pyrimidine bases, since they have a single ring of atoms. RNA has three of the same nucleotides, but instead of thymine, RNA has uracil, another pyrimidine base. RNA will come back later, but for the remainder of this section and the next few sections, we will discuss only DNA.

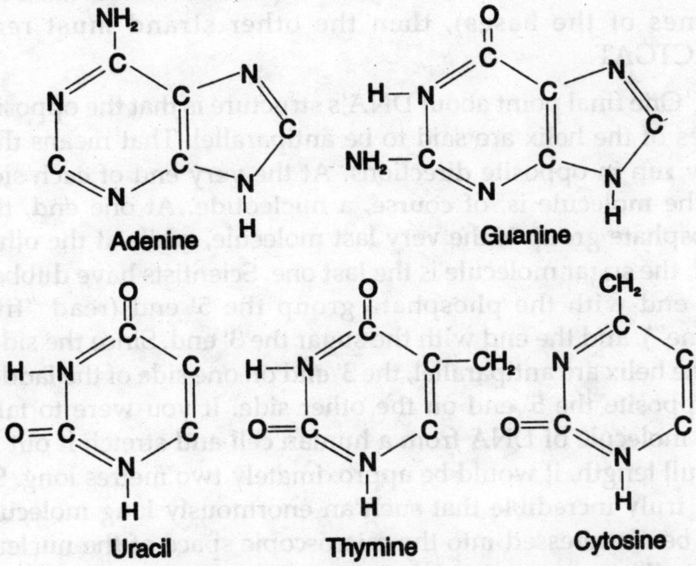

Adenine

Guanine

Uracil

Thymine

Cytosine

Knowing what DNA is composed of is only half of the mystery, as scientists still could not work out the physical structure of the molecule. In the 1940s, Erwin Chargaff made an important discovery which had significant implications regarding the structure of DNA. He found that a DNA molecule contains about the same amount of adenine as thymine, and about the same amount of cytosine as guanine. This countered an earlier suggestion that the four bases existed in equal amounts in the DNA molecule.

The now famous scientists who worked out the actual structure of DNA were James Watson and Francis Crick. Their work was published in 1953. Using data from many other scientists, they deduced that the DNA molecule was arranged in the form of a double helix; that is, it looks like a twisted ladder. The sides of the ladder are composed of nucleotides with their nitrogenous bases pointed toward the centre of the ladder. The rungs of the ladder are bonds between the bases on the opposite sides. To fit Chargaff's findings about the amounts of the bases in a DNA molecule, Watson and Crick suggested that adenine only forms a bond with thymine, and guanine only forms a bond with cytosine. So, if one side of the DNA molecule reads TTGACTA (we have abbreviated the names of the bases), then the other strand must read AACTGAT.

One final point about DNA's structure is that the opposite sides of the helix are said to be antiparallel. That means that they run in opposite directions. At the very end of each side of the molecule is, of course, a nucleotide. At one end, the phosphate group is the very last molecule, while at the other end, the sugar molecule is the last one. Scientists have dubbed the end with the phosphate group the 5' end (read "five prime"), and the end with the sugar the 3' end. Since the sides of the helix are antiparallel, the 3' end on one side of the ladder is opposite the 5' end on the other side. If you were to take one molecule of DNA from a human cell and stretch it out to its full length, it would be approximately two metres long. So it is truly incredible that such an enormously long molecule can be compressed into the microscopic space of the nucleus of a cell.

We will start from the very beginning of the packaging: the actual DNA molecule. It is first wrapped twice around a cluster of protein molecules called histones. This structure, a cluster of histones and two loops of DNA around it, is called a nucleosome. But this packing is not nearly enough to squeeze the tremendous DNA molecule into the nucleus. The nucleosomes are subsequently coiled together, and then this coil is arranged in tightly packed loops. This incredibly dense mass of loops and coils is the condensed chromatin that you would see in the nucleus of a cell.

DNA REPLICATION

Scientists knew that DNA had to replicate (copy) itself prior to cell division so that each daughter cell would obtain all of the genetic information. Based on the Watson-Crick model of DNA, three methods of DNA replication were suggested. They are called conservative, semiconservative, and dispersive replication.

In conservative replication, the double helix remains completely intact during the replication process, and an entirely new double helix is formed without destroying the original copy.

A second hypothesis is known as semiconservative replication, whereby the bonds between the bases are broken and the DNA molecule "unzips" into two strands. Alongside each of the two strands forms a new strand with the appropriate base pairs. Thus the final copies are half original DNA and half new DNA, in contrast to conservative replication in which the copies are either completely original or completely new.

The least likely candidate for DNA replication is dispersive replication. In this hypothesis, the DNA molecule is broken up into many small segments.

Alongside each segment forms an appropriate complementary segment, and then all of the segments are joined back together into two molecules of DNA. The final product is two strands of DNA with small pieces from the original DNA and small pieces from the new DNA.

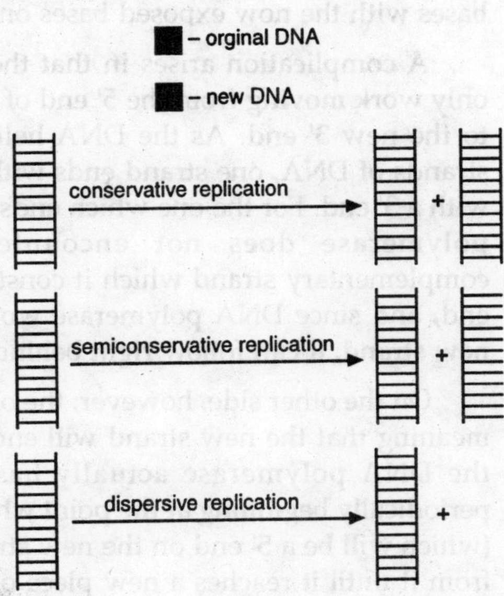

Through a series of experiments in 1958, two scientists, Matthew Meselson and Franklin Stahl, gathered evidence which disproved conservative and dispersive replication as possibilities for the method of DNA replication. Thus they provided strong evidence in support of semiconservative replication, which has since been accepted as the true method.

Of course, DNA replication is a bit more complex than a mere unzipping of the molecule and then copying each strand. A number of enzymes are involved in the process, and there are a few intricacies which one might not have originally expected. We will discuss the process in general first, and then we will discuss the difference between replication in prokaryotes and eukaryotes.

To begin the process of replication, an enzyme called DNA helicase moves down the DNA molecule and separates the two strands by breaking the bonds between the nitrogenous bases. This process begins at a point called the replication origin. Soon after the DNA helicase passes over a part of the DNA, an enzyme known as DNA polymerase bonds the appropriate bases with the now exposed bases on the single strand.

A complication arises in that the DNA polymerase can only work moving from the 5' end of the new DNA molecule to the new 3' end. As the DNA helicase separates the two strands of DNA, one strand ends with a 5' end and the other with a 3' end. For the one which ends with a 5' end, the DNA polymerase does not encounter a difficulty. The complementary strand which it constructs will end with a 3' end, and since DNA polymerase works from 5' to 3' on the new strand, it can follow right behind the DNA helicase.

On the other side, however, the old strand ends with a 3', meaning that the new strand will end with a 5'. In this case, the DNA polymerase actually has to work backwards, periodically beginning at the point where the DNA helicase is (which will be a 5' end on the new strand) and moving away from it until it reaches a new piece of DNA already formed. In this manner, the DNA polymerase creates small fragments of DNA called Okazaki fragments (named after Reiji Okazaki, who discovered them), which are then joined together by another enzyme called DNA ligase.

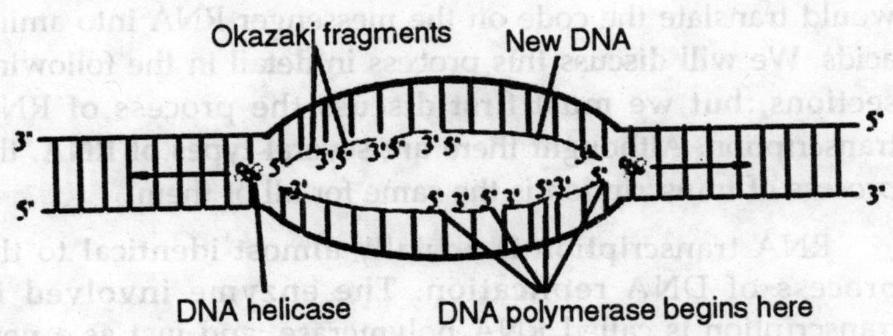

Prokaryotic DNA is arranged in a circular shape, and there is only one replication origin when replication starts. By contrast, eukaryotic DNA is linear; it does not connect end to end to form a circle. When it is replicated, there are as many as 1000 replication origins. Despite these differences, however, the underlying process of replication is the same for both prokaryotic and eukaryotic DNA.

TRANSCRIPTION OF RNA

Research by such scientists as George Beadle and Edward Tatum suggested that DNA in some way dictates the production of proteins in a cell. However, amino acid molecules cannot simply line up against a DNA molecule as the bases do in DNA replication; amino acids and nitrogenous bases are too unlike one another to possibly have a matching system. Therefore, biologists hypothesized that the ordering of the bases in a DNA molecule indirectly determined what proteins would be produced, and that certain specific patterns of bases would code for different amino acids which would then be bonded together to form proteins.

After a great deal of research, scientists determined the general method by which proteins were synthesized based upon the DNA code. First, a molecule of what is called messenger RNA would be synthesized through a process known as transcription based upon the nitrogenous bases in the DNA molecule. The messenger RNA would then move to ribosomes in the endoplasmic reticulum, where the ribosomes would translate the code on the messenger RNA into amino acids. We will discuss this process in detail in the following sections, but we must first discuss the process of RNA transcription. Althought there are several types of RNA, the process of transcription is the same for all of them.

RNA transcription is actually almost identical to the process of DNA replication. The enzyme involved in transcription is called RNA polymerase, and just as a new strand of DNA is formed by matching the corresponding base pairs along side of the original DNA, so an RNA molecule is synthesized by stringing together the appropriate bases as the DNA is unzipped. However, instead of thymine, RNA contains a different base, uracil. So, if a DNA strand read AGGTCG, the RNA strand produced by transcription would read UCCAGC.

You may be wondering how the RNA polymerase knows where to begin and end the transcription of RNA; after all, only certain parts of the DNA molecule need to be copied at certain times. The answer to this question is actually very satisfying; there is a certain sequence of bases on the DNA molecule called the promoter which the RNA polymerase enzyme must attach to in order to begin transcription. The enzyme moves down the DNA molecule, moving toward the DNA's 3' end, until it reaches a second sequence called the termination signal.

TYPES OF RNA AND THE M-RNA CODE

There are three types of RNA: messenger RNA (mRNA), transfer RNA (tRNA), and ribosomal RNA (rRNA). mRNA is used to carry the genetic code from the nucleus to the ribosomes in the endoplasmic reticulum, tRNA bonds to amino acids and it used in the synthesis of proteins, and rRNA is a component of ribosomes.

mRNA is by far the longest of the RNA molecules, since its function is to transport significant portions of the DNA code to the ribosomes. Before entering the cytoplasm on its way to the ribosomes, an mRNA molecule is modified somewhat. First, a 7-methylguanosine "head" is added which serves to help attach the mRNA to a ribosome during protein synthesis. Also, a so-called poly-A "tail" consisting of about 200 adenosine residues is attached to the end of the mRNA molecule. Research suggests that this tail protects the mRNA molecule from being destroyed by enzymes in the cytoplasm.

When the code on the mRNA is translated into a sequence of amino acids in order to form a protein molecule, tRNA serves as the molecule which transfers the amino acid to the ribosome when the codon requires it. On one end of the tRNA molecule is an area to which a specific amino acid attaches. On the other end is a sequence of three base pairs called the anticodon, the complement of the codon on the mRNA which codes for a certain protein. In the following section about a process called translation, the importance of this will become clear.

The third type of RNA is rRNA. The small subunit of a ribosome contains an rRNA strand 1542 nucleotides long, and the larger subunit contains two strands of rRNA, one 2904 nucleotides in length and the other 120. It is still not known how the rRNA and proteins in the ribosome fit together.

When scientists first began to tackle the problem of solving the genetic code, it was known that there are about 20 significant amino acids found in proteins. If each base coded for a particular amino acid, then only four amino acids would be possible. If instead two bases were required to code for a single amino acid, only 16 amino acids would be possible. Therefore, scientists hypothesized that three bases in sequence coded for one amino acid, since this would allow for the existence of 64 amino acids, well more than the 20 needed. This has since been proven to be true and each of the triplets has been termed a codon.

Through a great deal of experimentation, scientists were able to determine which codons on the mRNA molecule code for which amino acids. Most amino acids have several codons associated with them. The chart below shows all of the different amino acids and their codons.

Amino Acid	Associated Codon(s)
Alanine	GCA, GCC, GCG, GCU
Arginine	AGA, AGG, CGA, CGC, CGG, CGU
Asparagine	AAC, AAU, GAC, GAU
Cysteine	UGC, UGU
Glutamic acid	GAA, GAG
Glutamine	CAA, CAG
Glycine	GGA, GGC, GGG, GGU
Histidine	CAC, CAU
Isoleucine	AUA, AUC, AUU
Leucine	UUA, UUG, CUA, CUC, CUG, CUU
Lysine	AAA, AAG
Methionine	AUG
Phenylalanine	UUC, UUU
Proline	CCA, CCC, CCG, CCU
Serine	AGC, AGU, UCA, UCC, UCG, UCU
Threonine	ACA, ACC, ACG, ACU
Tryptophan	UGG
Valine	GUA, GUC, GUG, GUU
"Stop" codon	UAA, UAG, UGA

TRANSLATION

Translation is a term which refers to the actual synthesis of a protein molecule at the ribosomes based upon the codons in the mRNA. Once an mRNA molecule, created by transcription in the nucleus, reaches a ribosome, the small subunit of the ribosome attaches to the first codon (the one at the 5' end) on the mRNA molecule, which is always AUG. This codes for the amino acid methionine which subsequently becomes the first amino acid in the protein about to be produced.

After the small subunit has attached to the first codon, a tRNA molecule with the appropriate anticodon to match with AUG (the anticodon is UAC) bonds to the codon on the mRNA molecule. This tRNA molecule has attached to it a molecule of methionine. Next, the large ribosomal subunit attaches itself on top of the tRNA molecule. The large subunit has two sites at which tRNA molecules can be placed: the P site and the A site. When the large subunit attaches itself, the tRNA molecule with methionine is positioned in the P site.

The remaining steps of translation are repeated over and over again to create protein molecules. First, a tRNA molecule with the matching anticodon for the codon lined up with the A site attaches itself there. Then, a peptide bond forms between the two amino acids which are being held adjacent to one another. This bond formation requires energy from ATP. Once the bond is completed, the tRNA molecule in the P site releases its bond to its amino acid, and the tRNA in the A site moves to the P site. From this point, the process can be repeated to form long strings of amino acids.

When a so-called "stop" codon is positioned in the A site, a special releasing factor enters the A site, causing the entire amino acid chain to be released into the endoplasmic reticulum, eventually making its way to the Golgi bodies where it is packaged in a vesicle and sent to other parts of the cell.

MUTATIONS OF DNA

A mutation is a change in the DNA of an organism. Most mutations are harmful, but because of the redundancy of DNA (that certain codes are repeated over and over), a mutation in a single area may have only a minimal effect. We will discuss the two main types of mutations: point mutations and frame shift mutations.

A point mutation is very easy to understand. It simply means that one base is replaced with another. For example, if a part of the DNA should read TAC GGA ACT ATG but instead reads TAC GCA ATT ATG, then a point mutation has occurred in the third codon. Notice that all of the other codons remained the same. So if this part of the DNA were transcribed into mRNA which was later translated into a protein molecule, only one amino acid would be different than usual since only one codon was changed. Usually, a protein with only one amino acid different can still function normally.

By contrast to this minor effect, frame shift mutations can have much more serious consequences. A frame shift mutation occurs when a base is either added or deleted from the DNA sequence. Taking the example from before, suppose that a nucleotide with thymine was added in the second codon so that the sequence becomes TAC GGCA ACT ATG. Since the codons are triplets, this would actually be read as TAC GGC AAC TAT G... The first codon is the same, but all of the other ones are completely different, so the protein produced would also be very different from what it should be. Usually, the resulting protein will not be able to perform the function that the original one was meant to do. This is why frame shift mutations have far more severe effects than do point mutations.

Chapter 7

Unicellular Organisms and Plant World

Unicellular organisms are a very varied bunch. Some are smaller than others or are more complex, have more interesting names, or can live in very hostile environments. The purpose of this chapter is to discuss the many different types of unicellular organisms. We will cover the general method of classifying all organisms, then examine the major classes of prokaryotic organisms, followed by a discussion of eukaryotes. This chapter has a special focus on the information which may be useful for the Simulation. Although not all of the classes of unicellular organisms discussed here are used in the Simulation, all of them are significant.

The two important classes of unicellular organisms: prokaryotes and eukaryotes. Remember, eukaryotic cells are in general more complex than prokaryotic organisms. See this cell chart again:

	Prokaryote	Eukaryote
Cell membrane	yes	yes
Nucleus	no membranes	surrounded by two membranes
Endoplasmic reticulum	no	yes
Golgi bodies	no	yes
Mitochondria	no	yes
Vacuoles	no	yes
Lysosomes	no	yes
Ribosomes	yes	yes
Chromosomes	circular without histones	linear with histones

Organisms may also be classified by their method of nutrition and requirements for respiration. Those which produce their own food (through photosynthesis, for example) are called autotrophs. Heterotrophs are organisms which must ingest nutrients from the external environment. Cells which require oxygen to perform their nutritional process (whether it be photosynthesis or glycolysis) are called aerobes; those which do not need oxygen are called anaerobes.

Biologists have devised a system of classification for all life on earth based on assigning each organism to one of five large categories called kingdoms. The five kingdoms are Monera (prokaryotic organisms), Protista (single-celled eukaryotes), Fungi, Plantae, and Animalia. Only the first two kingdoms contain unicellular organisms, so we will discuss only those.

Within each kingdom are many additional categories called phyla (one such category is called a phylum). Organisms may be classified using additional subdivisions under the phyla, but for our purposes, we will not require this specificity.

All prokaryotic organisms are part of the kingdom Monera. As we discussed above, these organisms do not have membrane-bound organelles, and their DNA and other structures are chemically different from those in eukaryotic organisms. The kingdom Protista, on the other hand, contains unicellular eukaryotic organisms.

The remainder of this chapter is devoted to discussing each major type of unicellular organism with a special focus on information which may come in handy for the Simulation.

SPIROCHETES, MYXOBACTERIA, AND CYANOBACTERIA

Spirochetes

One important class of bacteria consists of spiral-shaped cells called spirochetes. Although these heterotrophs may be aerobic or anaerobic, they share the common distinct spiral feature.

They hold their shape through the means of special flagella called axial filaments which the cell wraps itself around, resulting in a spiral shape. In order to move, the axial filaments are rotated, the spiral turns, and the organisms swims along. Among the many varieties, one class of spirochetes is transmitted by fleas and causes Lyme disease.

Myxobacteria

The name myxobacteria is derived from the slimy mucus-like secretion which they release. The prefix myxo is equivalent to the word mucus; hence the name, myxobacteria. Most of them live in soil rather than an aquatic environment. They do not possess flagella but are able to glide through the use of strange fibrils on the inside of the cell. Myxobacteria are heterotrophic aerobic organisms.

One of the most interesting aspects of myxobacteria is their method of reproduction. Unlike most prokaryotes which reproduce simply by division, myxobacteria first form large structures called fruiting bodies which may have between thousands and millions of organisms. Some of the cells in the fruiting body divide, and the daughter cells come together to form what are called cysts, small groups of cells which are released into the environment. When a cyst finds an environmentally-friendly location, the cells divide and grow.

Cyanobacteria

Cyanobacteria are prokaryotic organisms sometimes referred to as blue-green algae. They are autotrophic and utilize photosynthesis, using a type of chlorophyll called chlorophyll a. Cyanobacteria, which can be aerobic or anaerobic, are in many ways the most advanced class of prokaryotes in that they possess specialized membranes which help perform specific functions, especially those devoted to photosynthesis. However, they do not possess flagella like many other prokaryotes do.

One interesting aspect of cyanobacteria is that they often form sophisticated colonies in which different cells actually perform certain roles. These colonies usually can form in any wet environment, including the ocean and freshwater lakes.

Algae

Algae is a term which is no longer used to officially classify organisms, but it is still informally used to refer to a group of organisms in the kingdom Protista. They possess all of the important characteristics of eukaryotic organisms; they have specialized, membrane-bound organelles and a DNA structure more complex than prokaryotes. In addition, most algae are photosynthetic, have a cell wall, and are found near the surface of bodies of water where light is plentiful. They can also form colonies. Algae may be multicellular, but we will concern ourselves only with the unicellular varieties. The first division of algae which we will discuss is found in the next section: the euglenas.

Euglenas

Euglenas are one of the most well known types of unicellular organisms. As a type of algae, they have chloroplasts to regulate the process of photosynthesis. Although they perform photosynthesis like most autotrophic cells, they do not have a cell wall and can actually survive by ingesting food like a heterotroph if there is not sufficient light to drive photosynthesis. As a result, many biologists view the euglena as the ancestor of the cells of both plants and animals.

One important structure unique to euglenas is known as the stigma, a primitive but effective organelle which is sensitive to light. The euglena moves (using its long flagellum) toward light detected by the stigma. In this manner, the euglena can seek out the light needed for photosynthesis. They also have large structures called pyrenoid bodies which store starch in a form called paramylum as a spare energy source.

Euglenas also possess contractile vacuoles to force out excess water which is diffused into the cell; otherwise, the cell would burst as the water rushes in. This is especially important since euglenas usually live in freshwater environments where there is little salt in the water. Thus the water tends to diffuse into the cell faster than it would in a saltwater environment like the ocean.

Euglenas have two methods of locomotion. They can use their long flagellum, which whips back and forth and propels the euglena, or a structure known as the pellicle. The pellicle, which is made of protein, lies just next to the cell membrane. Its "wiggling" motion can also be used to move the euglena.

Euglenas reproduce asexually by dividing longitudinally (lengthwise). However, their method of asexual reproduction is a bit "sloppy" in that the movement of chromosomes during division is more irregular than that in other cells which undergo mitosis. Euglenas are not known to reproduce sexually.

CHRYSOPHYTES AND PYRROPHYTES

Chrysophytes

Chrysophytes are another division of algae. They contain chlorophyll but also possess a carotenoid (pigment) called fucoxanthin which gives them a yellow-brown colour. Unlike the euglenas, chrysophytes have rigid cell walls which are often strengthened by silicon compounds. An additional distinction between chrysophytes and euglenas is that chrysophytes store their food as oils, whereas euglenas utilize starch. Many varieties of chrysophytes have flagella, although there are those which do not. They usually live in marine (ocean) environments. Chrysophytes generally reproduce asexually, although some types reproduce sexually through a process called meiosis.

Pyrrophytes

The third and final class of unicellular algae which we will discuss is the division pyrrophyta. Pyrrophytes, also known as dinoflagellates, have a rigid cell wall made of cellulose unlike the cell walls of chrysophytes which contain silicon compounds. They usually have two flagella. In addition, pyrrophytes are also notable in that they are bioluminescent; that is, they naturally glow in the dark!

Pyrrophytes store their food as both starch and oil, whereas euglenas and chrysophytes utilize only one of these methods. They are usually found in marine environments, and they usually reproduce asexually through mitosis.

Protozoa

Just as algae refers to the autotrophic protists, a second informal term, protozoans, refers to heterotrophic members of the kingdom Protista. Most protozoans reproduce asexually, although some take part in complex sexual cycles of reproduction. The following sections describe four significant phyla of protozoans: Mastigophora, Sarcodina, Sporozoa, and Ciliophora.

PHYLA MASTIGOPHORA AND SARCODINA

Phylum Mastigophora

Mastigophores are the most primitive type of protozoans. They often have many flagella, and some are able to form pseudopodia. As discussed in Chapter Three, pseudopodia are finger-like extensions of the cytoplasm which can be used to surround engulf food particles or for movement. Mastigophores usually reproduce asexually through mitosis, although some varieties can reproduce sexually. They are usually parasitic; that is, they live inside another organism (the host organism) to obtain nutrients and in effect harm the host organism. However, some mastigophores are free-living, but since most are not, they are not used in the Simulation.

Phylum Sarcodina

The sarcodines are a much more familiar group of protists than the mastigophores.

The most well-known example of a sarcodine is the famous amoeba. Lacking any rigid structure outside of their cell membrane, sarcodines can freely change their shape and form pseudopodia. Sarcodines can live in both freshwater and marine environments. They can reproduce both asexually and sexually, and they are usually free-living. Like mastigophores, sarcodines use pseudopodia to move and capture food.

Although the amoeba is generally thought of as lacking any structure, some have shells, and most other types of sarcodines also have shells. One class of sarcodines, the foraminiferans, possess calcareous shells (they are made of $CaCO_3$). Radiolarians also have shells, but theirs contain silica. While the latter two groups usually live in saltwater environments, the heliozoans live in freshwater. They too can have shells which contain silica.

PHYLA SPOROZOA AND CILIOPHORA

Phylum Sporozoa

Sporozoans are parasites; they live at the expense of another organism. One of their significant characteristics is that they lack cilia and flagella. They generally obtain nutrients by absorbing organic molecules from the host organism.

Sporozoans often have very complicated life cycles. Take, for example, the Plasmodium vivax, the organisms responsible for the disease malaria. It grows inside of a mosquito's stomach, and, after reaching maturity, migrates to the mosquito's salivary glands where it releases thousands of small cells called sporozoites. When the mosquito bites a person, the sporozoites are transmitted into the human bloodstream. They enter the liver where they divide and enter red blood cells. Eventually, they break free from the red blood cells as gametes (causing the fever associated with malaria). If a mosquito bites the person at this time, the gametes can be ingested into the mosquito's stomach where they unite and grow into new organisms.

Phylum Ciliophora

As their name suggests, must members of the phylum Ciliophora (they are called ciliates) have great amounts of cilia. Ciliates may also have structures called trichocysts, organelles which can be discharged from the cell. Trichocysts may be used to anchor the organism or to capture prey by paralyzing it with a trichocyst with a poisoned tip.

Ciliates, which can live in both freshwater and marine environments, usually consume bacteria or other protists. They also feed on organic material which may be floating in the surrounding water.

An interesting aspect of ciliates is their possession of two nuclei. One is called the micronucleus, and the other is known as the macronucleus. It is currently believed that the micronucleus is involved with sexual reproduction, while the macronucleus controls such aspects of the organism's life as growth, respiration, and asexual reproduction.

The most well known ciliate is the paramecium. The body of a paramecia is completely covered by cilia, allowing for efficient locomotion. They have a gullet (also known as an oral groove) through which large food particles may pass. Smaller particles are consumed through the cell membrane into vacuoles by phagocytosis. Most paramecia also have a contractile vacuole. Paramecia may reproduce asexually through mitotic division, but they may also engage in a sharing of genetic information with another paramecium through a process called conjugation. In this process, the two paramecia align their gullets and allow material in the nucleus to pass through. In this manner, they keep some of their initial DNA but obtain new DNA, so that new paramecia formed through mitosis later on may have slightly varying characteristics.

WORLD OF PLANTS

CHARACTERISTICS AND STRUCTURES OF PLANTS

Plant Characteristics

Photosynthesis: Plants obtain their energy by converting light energy into chemical energy by means of photosynthesis, which involves two basic sets of reactions that take place in the chloroplast. The light dependent reactions involve the absorption of light energy by specialized pigments including chlorophyll, and the reaction within the thylakoid. The energy is absorbed by electrons and used to create ATP and a compound called NADPH provides the hydrogen for the subsequent reaction. These compounds carry energy and hydrogen ions (in the case of NADPH) into the light independent reactions. These reactions bring carbon dioxide into the plant's metabolism. Eventually, all these materials are combined into glucose and other carbohydrates. In general, the plant uses its root to extract water in the soil, its leaf to absorb carbon dioxide from the air,

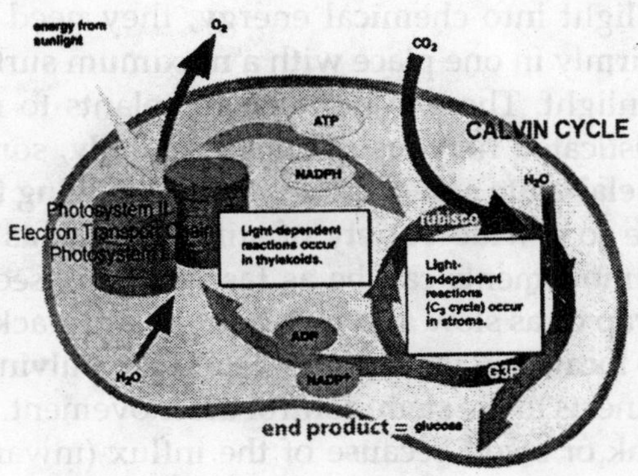

Fig. Pic: Photosynthesis, energy from the sun, and the stem for transportation of water and nutrients. Oxygen is a by product exits through the leaf.

Leaves are the organs of photosynthesis in vascular plants. A leaf usually consists of a flattened blade and a petiole, which connects the blade to the stem. The blade may be single or it may be composed of several leaflets. Externally, it is possible to see the pattern of the leaf veins that bring water to the leaf; they are the final extensions of the vascular tissue. The cross section of a typical leaf, its external and internal structures are shown in figure. At the top and bottom of the leaf is a layer of epidermal tissue that often bears protective hairs and glands that produce irritating substances to discourage herbivores. The epidermis is covered by a waxy cuticle that keeps the leaf from drying out. Unfortunately, it also prevents gas exchange because the cuticle is not gas permeable. However, the epidermis, particularly the lower one, contains openings called stomata that allow gases to move into and out of the leaf. Each stoma has two guard cells that regulate its opening and closing. The body of a leaf is composed of mesophyll tissue, which contain many chloroplasts and carry on most of the photosynthesis for the plant. Leaf veins consist of a strand of xylem and a strand of phloem surrounded by a bundle sheath. The bundle sheath cells in C_3 plants do not contain chloroplasts; while the C_4 plants characteristically have chloroplasts and are more efficient in fixing carbon dioxide.

Locomotion: It is often said that since plants simply transform light into chemical energy, they need only to be anchored firmly in one place with a maximum surface area to capture sunlight. There is no need for plants to move or to have sophisticated nervous systems. Actually, some of them have quite elaborate and creative ways of moving their leaves in response to a wide variety of stimuli, such as touch and light. Leaf movements can be as fast as a millisecond in the Venus flytrap or as slow as a half-hour in sun-tracking plants. Motor cells located in the region—called the pulvinus—where the leaf connects to the stem, control the movement. These cells either shrink or swell because of the influx (inward flow) or efflux (outward flow) of water. For example, if the cells at the top of the pulvinus swell and the bottom cells shrink, the leaf tilts downward. The changes in cell volume are a direct result of changes in the concentration of certain ions in the cell, such as potassium and chloride.

For example, when a large amount of potassium enters the cell through special potassium channels, a large amount of water must enter so that the concentration of potassium stays constant. It is unclear what is the signaling mechanism to open the ion channels. Different plants have evolved different signaling mechanisms depending on what triggers the leaves to move.

Fig. Venus Flytrap

Reproduction and Growth: Asexual reproduction is very common among plants, while sexuality is not well-defined (90% have both male and female parts). Plant cells are rarely terminally differentiated. Since most cells are totipotent, detached limb can be re-generated readily. Plants do not establish a germ line (special cells destined to produce gametes). Higher plants have sex organs with an outer layer of nonreproductive cells that can prevent desiccation of gametes - the flower. The developing diploid embryo is protected from drying out by providing it with water and nutrients within the female reproductive structure - the fruit. Growth in plants can be indeterminate. All plants have a two-generation life cycle known as alternation of generations. This means that a plant exists in 2 forms: the haploid generation is the gametophyte that produces gametes; and the diploid generation is the sporophyte that produces spores by meiosis. Spores are haploid structures that develop or mature into the gametophyte plant. Usually, lower plants spend their life longer in haploid (1N) generation.

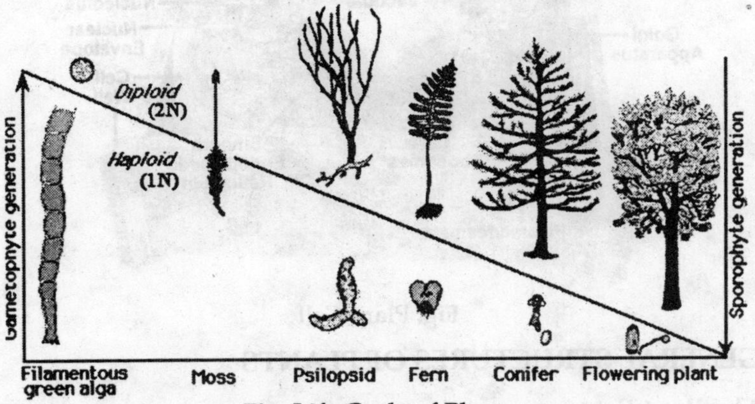

Fig. Life Cycle of Plants

In addition to a cell membrane, plant cells are surrounded by a cell wall that varies in thickness depending on the function of the cell. All plant cells have a primary cell wall, having as its main constituent cellulose molecules united into threadlike microfibrils. Several microfibrils, in turn, are found in fibrils, and within the wall there are layers of fibrils lying at right angles to one another for added strength . Some cells in woody plants have a secondary cell wall that forms inside the primary cell wall. Secondary cells contain lignin (major component of wood), a substance that makes secondary cell walls even stronger than primary cell walls.

Plant support cells have secondary walls; the cell dies and the strong wall remains as supporting material. The plasmodesmata is a very fine thread of cytoplasm that passes through openings in the walls of adjacent cells and forms a living bridge between them.

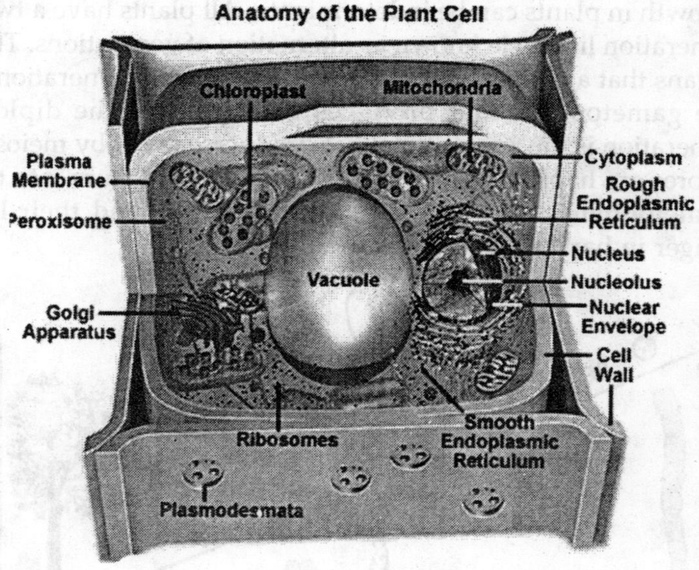

Fig. Plant Cell

GENERAL STRUCTURES OF PLANTS

The Root System

- It is usually underground.
- It anchors the plant in the soil.
- The root absorbs then conducts water and minerals to the stem.
- It is a food storage.

The Shoot System

- **It is usually above ground to elevates the plant from the soil.**
- **The shoot system includes the stem, the leaves, and the reproductive organs.**
- **It provides many functions including -**
  - **photosynthesis.**
  - **reproduction and dispersal.**
  - **nutrients and water conduction.**

Algae and Classification of Plants

The term algae is used for aquatic unicelluar organisms that photosynthesize as do terrestrial plants. All algae contain green chlorophyll, but they also can contain other pigments that mask the colour of the chlorophyll. The chlorophylls selecting different part of the spectrum are separated into type a, b, c, and d. Thus there are green, golden brown, brown, and red algae. Green algae can be single-celled, colonial, filamentous, and multicellular. It is believed that the green algae are ancestral to the first plants because both of these groups possess chlorophylls a and b, both store reserve food as starch, and both have cell walls that contain cellulose. Seaweeds such as Ulva are multicellular algae carrying chlorophyll. They are anchored firmly to the rock by holdfasts. Their appearance give a false impression that they are plants having root, stem, and leaf.

While a few of the algae such as some species of Fucus follow the diplontic lifecycle, most of them such as the green algae Chlamydomonas spend most of their life in the haploid generation. Usually, this protist practices asexual reproduction, and the adult divides to give zoospores that resemble the parent cell. During sexual reproduction, gametes of two different strains come into contact and join to form a zygote. A heavy wall forms around the zygote, and it becomes a zygospore. The zygospore is able to survive until conditions are favorable for germination and subsequent production of 4 zoospores by meiosis. The gametes are isogametes; that is, they look exactly alike. Sexual reproduction aids the process of evolution because it offers means to produce variations in addition to mutations. Note that embryo is absent in this kind of lifecycle.

Table 01 below classifies the plants according to their characteristics. Algae are included because they satisfy the most basic definition of a plant - the use of chlorophyll to produce energy. The table shows that the properties are acquired gradually by each group as they progress from algae to the flowering plants.

Phylum	Chl. a & b	Embryos	Stele	Seed	Flower	# of Living Sp.
Algae	Yes	No	No	No	No	14000
Bryophytes	Yes	Yes	No	No	No	22500
Pteridophytes	Yes	Yes	Yes	No	No	10800
Gymnosperms	Yes	Yes	Yes	Yes	No	680
Angiosperms	Yes	Yes	Yes	Yes	Yes	260000

- Table 01 Classification of Plants
- Stele is the cylindrical bundle in the stems and roots = vascular tissue.

BRYOPHYTES (MOSS, LIVERWORT)

The bryophytes include liverworts and mosses. Most species of liverworts are "leafy" and look somewhat like mosses, but close examination shows that the body of a liverwort has distinct top and bottom surfaces, with numerous rhizoids (rootlike hairs) projecting into the soil. In contrast, a moss has a stemlike structure with radially arranged, leaflike structures. Rhizoids anchor the plant and absorb minerals and water from the soil. Because bryophytes do not have vascular tissue, they lack true roots, stems, and leaves. Instead, they have rhizoids, stemlike and leaflike structures.

In mosses, the gametophyte is dominant - it is longer lasting. In some mosses, there are separate male and female gametophytes. At the tip of a male gametophyte are antheridia, in which swimming sperms are produced. After rain or heavy dew, the sperm swim to the tip of a female gametophyte, where eggs have been produced within the archegonia.

Antheridia and archegonia are both multicellular structures, and each has an outer layer of jacket cells that protects the enclosed gametes from drying out. After an egg is fertilized, the developing sporophyte is retained within the archegonium as an embryo. The sporophyte, which is dependent on the gametophyte, consists of a foot that grows down into the gametophyte tissue, a stalk (seta), and an upper capsule, or sporangium, where meiosis occurs and where haploid spores are produced. In some species of mosses, a hoodlike covering is carried upward by the growing sporophyte. When this covering and the capsule lid falloff, the spores are mature and ready to escape. The rings of "teeth" projected inward from the edge of the capsule allows spores to be released only at times when the weather is dry (when they are most likely to be dispersed by wind). When a spore lands on an appropriate site, it germinates. The single row of cells that first appears branches, giving an algalike sturcture called a protonoma. After about three days of favorable growing conditions, new moss plants appear at intervals along the protonema. Each of these consists of the rootlike rhizoids and the upright shoots of a moss gametophyte. The gametophytes produce gametes, and the moss life cycle begins again.

Sperms are released when the antheridium ruptures, thus allowing them to swim freely in a water film toward the archegonium. The zygote is the first cell of the new sporophyte just after fertilization. The zygote divides by mitosis into a multicellular embryo within the archegonium. This is the crucial step that separates plants from algae. The embryo then inserts an absorbing organ called the foot into the female stem tip. The other end of the embryo grows up and above the female stem to form a stalk (seta) and sporangium (capsule) anchored in the old archegonium. Early in its development the sporophyte is typically green, but by the time it is mature it is usually non-photosynthetic and dependent on the gametophyte for water and nutrients. Within the sporangium special cells called sporocytes divide by meiosis to produce thick-walled haploid spores. In the more advanced plant species, the embryo is enclosed within the seed.

PTERIDOPHYTES (FERN, CLUB MOSS, HORSETAIL)

Vascular plants (also called tracheophytes) are believed to have evolved sometime during the late Silurian Period. The primitive vascular plants include the whisk ferns (Psilopsid), the club mosses, and the horsetails. The whisk ferns is of particular interest because they may be the most primitive. It bears considerable resemblance to the extinct rhyniophytes. Its sporophyte consists of stems with scalelike structures but no leaves. There is a horizontal stem (lacking roots), from which rhizoids grow, and there are green, photosynthetic, upright branches with tiny, scalelike structures that grow upward. Sporangia are located on the branches. The gametophyte is separate from and smaller than the sporophyte; it also lacks vascular tissue.

In general, the tracheophytes have two types of vascular tissue. Xylem conducts water and minerals up from the soil, and phloem transports organic nutrients from one part of the body to another. Because they have vascular tissue, the specialized body parts of tracheophytes can be called properly roots, stems and leaves.

Pic: Ferns

The life cycle of a common fern of the temperate zone. Young fronds grow in a curled-up form called fiddleheads, which unroll as they grow. The fronds often are subdivided into a large number of leaflets. The sporophyte fern plant represents the dominant generation. Sporangia develop in clusters called sori, which are protected by a covering, the indusium (not shown). Within the sporangia, meiosis occurs and spores and produced. The gametophyte is a tiny (1-2 cm), heart-shaped structure called a prothallus. The antheridia and archegonia develop on the under side of a prothallus. Fertilization takes place when moisture is present because the spiral-shaped sperm must swim from the antheridia to the archegonia. The resulting zygote soon develops into a sporophyte embryo consisting of a foot, a root, a stem, and a leaf. The root grows down into the soil, and the frond grows upward through the prothallus notch. As the sporophyte matures, the prothallus shrivels and disappears. Since the gametophyte lacks vascular tissue, and the swimming sperm relies on moisture to approach the egg, ferns are likely to be found in habitats that are at least seasonally moist. Once established, the sporophyte of some ferns can spread by vegetative reproduction into drier areas because this generation has vascular tissue.

The vascular structures in the ferns are primitive in comparison to the more advanced plant species. They have the rhizome, which can be compared to the stem of a flowering plant. In many cases the rhizome can be inconspicuous or even entirely

underground. Rhizomes of tree ferns on the other hand may be 60 cm in diameter and up to 12 metres tall. The fronds (leaves) arise from the upper side or in one or more rows laterally on each side from the rhyzome. They are composed of two main structures: the stipe (stalk) and the blade (the leafy outcroppings). Roots are formed from the rhizomes or sometimes from the stipe. The roots usually do not divide once they grow from the rhizome. Tree fern roots grow down from the crown and help thicken and strengthen the trunk . The roots anchor the plant to the ground and absorb water and minerals. The internal structures of the rhizome, the root, and the leaf an be seen in plants anatomy.

In the more advanced plant species, the outermost tissue of the stem is the epidermis. The stem has distinctive vascular bundles, where xylem and phloem are found. In each bundle, xylem is typically found toward the inside and phloem is toward the outside. In the dicot stem, the bundles are arranged in a distinct ring that separates the cortex from the central pith. The cortex is sometimes green and carries on photosynthesis, and the pith may function as a storage site for the products of photosynthesis. In the monocot stem, the vascular bundles are scattered throughout the stem, and there is no well-defined pith. It is similar to the more primitive type. Secondary growth of stems is seen primarily in woody plants, such as trees that live for many years. Almost all trees are dicots. Primary growth in woody plants occurs for a short distance beneath the apical meristem. Secondary growth occurs in the vascular and cork cambia. Vascular cambium begins as meristematic cells between the xylem and the phloem of each vascular bundle. Then these cells, join to form a ring of meristematic tissue adding to the girth of the stem. Cork cambium is located beneath the epidermis. It produces tissue that disrupts and replaces the epidermis with cork cells, which are impregnated with suberin (a waterproof substance). Dead cork allows gas exchange in pockets of loosely arranged cells, called lenticels. A woody stem has three distinct areas: the bark (containing cork, cork cambium, cortex, and phloem), the wood, and the pith. In large trees, only the more recently formed layer of xylem, the sapwood, functions in water transport. The older inner part, called the heartwood, becomes plugged with deposits, such as resins, gums, and other substances.

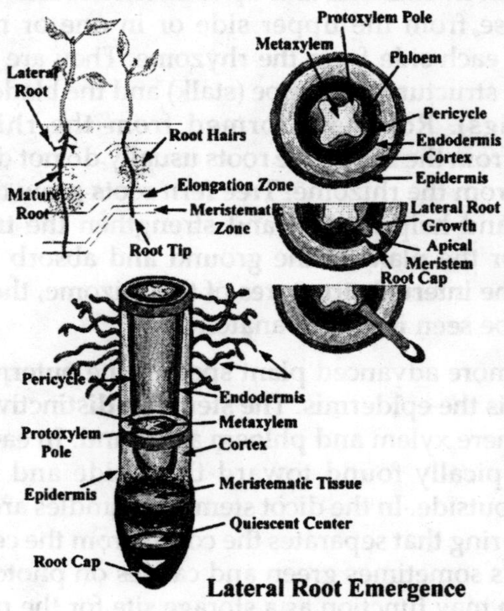

Lateral Root Emergence

Pic: Stem Anatomy

Figure above depicts a longitudinal section of a root. At the bottom is an area of cells called the root cap, a thimble-shaped mass of parenchymal cells (relatively unspecialized cells) that is a protective covering for the root tip, and the cells in the next region – the region of cell division. Cells in the root cap have to be replaced constantly because they are ground off as the root pushes through abrasive soil particles. The next area - the zone of cell division - is the area where new cells are continually being formed through repeated cell divisions. These cells are thin-walled and easily ruptured by soil particles were it not for the root cap's protection. Next is the zone of cell elongation. Here the cells take up large amounts of water and increase in volume. The increase in cell volume of these cells is primarily responsible for pushing the root through the soil. The next zone is the zone of cell maturation and differentiation. The fully elongated cells in these zones matured and began differentiating into various tissues such as the xylem, phloem, pith, cortex, and others. This zone, the zone of maturation and differentiation begins where the root hairs first become evident. These root hairs are only extensions of the epidermal cells - as may be seen in the inset drawing down. Branch roots have formed beyond these zones.

The absorbed water and minerals pass through the cortex, a tissue composed of parenchymal cells. The water and minerals are forced by a strip of waxy material (the Casparian strip) in the endodermis to move one way into the vascular cylinder. Within the vascular cylinder, water and minerals are transported upward by way of the xylem and the products of photosynthesis most often are transported downward by way of the phloem for storage in the cortex. Lying between the endodermis and the vascular tissue is the pericycle, composed of parenchymal cells, that retains the ability to undergo cell division and on occasion produces branch roots. The pericycle alos contributes to the formation of vascular cambium, whicn is meristematic tissue lying between xylem and phloem that is capable of producing new vascular tissue. Monocot roots often have pith, which is centrally located ground tissue. In a monocot root, pith is surrounded by a ring of alternating xylem and phloem bundles. They also have pericycle, endodermis, cortex, and epidermis.

Gymnosperms (Cyced, Ginko, Conifers)

The gymnosperms produce naked seeds; that is, the seeds are not enclosed by fruit. There are four divisions of gymnosperms (Cycadophyta, Ginkgophyta, Gnetophyta, and Coniferophyta) Cycads are cone-bearing, palmlike plants found today mainly in tropical and subtropical regions. Only one species of ginkgo, the maidenhair tree, survives today. The gnetophyta has only three genera left. The largest group of gymnosperms is the cone-bearing conifers, which include pine, cedar, spruce, fir, and redwood trees. These trees have needlelike leaves that are well adapted to not only hot summers but also cold winters and high winds. Most gymnosperms are evergreen trees.

The sporophyte is dominant in the pine life cycle. Typically, the male pine cones are quite small and develop near the tips of lower branches. Each scale of the male cone has two or more microsporangia on the underside. Inside the microsporangia are microspore mother cells that undergo meiosis and develop into mature pollen grain (with two lobular wings), which is a sperm-bearing male gemetophyte. The female pine cones are larger and located near the top of the tree.

Each scale of the female cone has two ovules that lie on the upper surface. Within the ovule, a megaspore mother cell undergoes meiosis and develops into mature female gametophyte, which has 2 - 6 archegonia, each containing a single, large egg lying near the ovule opening. During pollination, pollen grains are transferred from the male cone to the female cone. Once enclosed within the female cone, the pollen grain develops a pollen tube that slowly grows toward the ovule.

The pollen tube discharges two nonflagellated sperms. Only one of the sperms fertilizes an egg in the ovule 15 months after pollination. After fertilization, the ovule matures and becomes the seed composed of the embryo, its stored food, and a seed coat. Finally, in the third season, the female cone, by now woody and hard, opens to release its seeds, whose wings are formed from a thin, membranous layer of the cone scale. When a seed germinates, the sporophyte embryo develops into a new pine tree, and the cycle is complete.

The outermost layer of the conifer seed is the seed coat. It originates from the mothe tree and is diploid. The seed coat has three layers: the outer layer; the thicker, tough stony or middle layer; and the inner layer. Some species of conifers have resin vesicles in the middle or outer layers of the seed coat. These resin vesicles may play a role in seed coat dormancy, protecting dehydration, and deterring seed herbivory. Immediately inside the cell wall is the nucellus, a papery layer surrounding the megaspore cell wall.

Inside the megaspore cell wall is the megagametophyte, the haploid nutritional tissue found in gymnosperm seeds. Early in the development of cones, megagametophytes produce egg cells which are fertilized by male gemetes to produce zygotes. Zygotes develop into embryos. The megagametophyte then plays its second functional role, surrounding the embryo, protecting and nourishing it. The megagametophyte in Douglas-fir is 60% lipids, 16% proteins, and 2% sugars, making it a high-energy and nutrient tissue both for the embryos it contains and for a plethora of seed predators ,including small mammals, birds, and insects. The embryo is found in the corrosion cavity, a pit in the centre of the megagametophyte that is fully filled by the embryo in mature seeds.

It consists of the cotyledons (first leaf), shoot apical meristem, root apical meristem, root cap and suspensor. The cotyledons and shoot apical meristem point towards the wider end of the seed; while the radicale (embryonic root) and suspensor are at the more pointed end. The suspensor is found at the base of the root cap and plays a role early in embryo development by pushing the embryo into the megagametophyte.

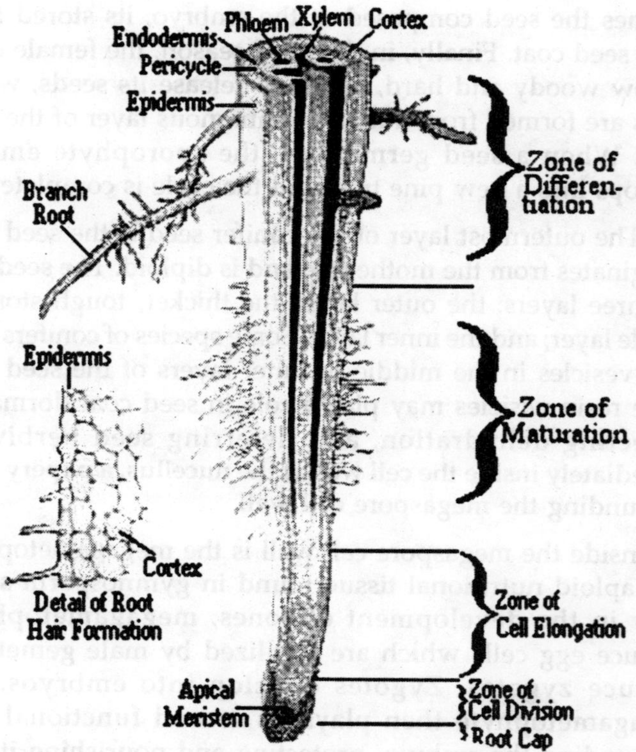

Fig. Anatomy

ANGIOSPERMS (OAK, MAPLE, BASIL,)

- Angiosperms are the flowering plants with seeds enclosed by fruit. All hardwood trees (broad-leaved trees, e.g., oak, ...), including all the deciduous trees (trees that shed their leaves in the fall, e.g., maple, ...) of the temperate zone and the broad-

leave evergreen trees (e.g., boxwood, ...) of the tropical zone, are angiosperms, although sometimes the flowers are inconspicuous. All herbaceous (nonwoody, nonpersistent) plants (e.g., basil, ...) common to our everyday experience, such as grasses and most garden plants are flowering plants Angiosperms are adapted to every type of habitat, including water (e.g., water lilies, ...). Angiosperms have well-developed vascular and supporting tissues. Their xylem tissue contains vessel elements beside the tracheids. Thus the woody angiosperms are considered as hardwood trees, whereas the gymnosperms are softwood trees.

The angiosperms are divided into two classes: the monocots (e.g., rice, ...) and the dicots (e.g., potato, ...). The distinction between these two groups is not always clear, some of the general characteristics (including the gymnosperms) are outlined in Table 02 below.

Table 02 Monocots, Dicots, and Gymnosperms Comparison

Characteristic	Dicots	Monocots	Gymnosperms
Embryo	Two cotyledons (seed leaves)	One cotyledon (seed leaf)	One to many
Flowers	Parts in 4/5	Parts in 3x	No true flower
Vascular Bundles	Ring	Scattered	Ring
Habit	Herbaceous or Woody	Herbaceous	Herbaceous or Woody
Roots	Taproot	Fibrous	Taproot
Leaf Venation	Net	Parallel	Needle-like
Pollen	Tricoplate (3 furrows or pores)	Monocoplate (1 furrow or pore)	Tow lobular wings
Growth	Primary and Secondary	Primary	Primary and Secondary

In angiosperms, the reproductive structures are located in the flower. The flower attracts insects and birds that aid in pollination, and it produces seeds enclosed by fruit. There are many different types of fruits, some of which are fleshy (e.g., , apple, tomato, peach, ...) and some of which are dry (e.g., pea enclosed by pod, nut, grain, ...). They all provide protection for the seeds. The life cycle of the flowering plant is shown in figure . Within a flower, there is a diploid megaspore mother cell in each ovule of the ovary. The mother cell undergoes meiosis, producing one functional megaspore, whose nucleus divides mitotically until there are eight haploid nuclei. This is the female gemetophyte, which sometimes is called the embryo sac. At one end of the embryo sac there the three cells, one of which is the egg cell. Male gametophytes are produced in the stamens. An anther contains four pollen sacs with many microspore mother cells, each of which undergoes meiosis to four microspores. After a mitotic division, each misrospore has two cells, one of which later divides again to give two sperm. Pollination, which is simply the transfer of pollen from the anther to the stigma, is brought about by wind or with the assistance of a particular pollinator. The plant uses the pollinator to ensure cross-pollination, and the pollinator uses the plant as a source of food in the form of nectar. When a pollen grain lands on a stigma of the same species, it germinates, forming a pollen tube. The pollen tube grows as it passes between the cells of the stigma and the style to reach the female gemetophyte.

Double fertilization takes place to produce seeds and fruits. One sperm nucleus from the pollen tube unites with the egg nucleus, forming a zygote, and the other sperm nucleus unites with the polar nuclei, forming a triploid (3N) endosperm nucleus. The endosperm nucleus divides, forming the endosperm, which is a nutrient material for the developing embryo and sometimes for the young seedling as well. The zygote develops into an embryo. The outer layers (integuments) of the ovule harden and become the seed coat. A seed is a structure formed by the maturation of the ovule; it contains a sporophyte embryo plus stored food. The ovary and sometimes other floral parts develop into the fruit. A fruit is a mature ovary that usually contains seeds. Therefore, angiosperms are said to have covered seeds.

Chapter 8

Ecology

Ecology is the scientific study of the distribution and abundance of living organisms and how the distribution and abundance are affected by interactions between the organisms and their environment. The environment of an organism includes both physical properties, which can be described as the sum of local abiotic factors such as insolation (sunlight), climate, and geology, as well as the other organisms that share its habitat. The term *oekologie* was coined in 1866 by the German biologist Ernst Haeckel the word is derived from the Greek (*oikos*, "household") and *logos*, "study"); therefore "ecology" means the "study of the household [of nature]". The word "ecology" is often used in common parlance as a synonym for the natural environment or environmentalism. Likewise "ecologic" or "ecological" is often taken in the sense of environmentally friendly.

SCOPE

Ecology is usually considered a branch of biology, the general science that studies living organisms. Organisms can be studied at many different levels, from proteins and nucleic acids (in biochemistry and molecular biology), to cells (in cellular biology), to individuals (in botany, zoology, and other similar disciplines), and finally at the level of populations, communities, and ecosystems, to the biosphere as a whole; these latter strata are the primary subjects of ecological inquiries. Ecology is a multi-disciplinary science. Because of its focus on the higher levels of the organization of life on earth and on the interrelations between organisms and their environment, ecology draws heavily on many other branches of science, especially geology and geography, meteorology, pedology, genetics, chemistry, and physics. Thus, ecology is considered by some to be a holistic science, one that over-arches older disciplines such as biology which in this view become sub-disciplines contributing to ecological knowledge.

Agriculture, fisheries, forestry, medicine and urban development are among human activities that would fall within Krebs' (1972: 4) explanation of his definition of ecology: "where organisms are found, how many occur there, and why".

As a scientific discipline, ecology does not dictate what is "right" or "wrong". However, ecological knowledge such as the quantification of biodiversity and population dynamics have provided a scientific basis for expressing the aims of environmentalism and evaluating its goals and policies. Additionally, a holistic view of nature is stressed in both ecology and environmentalism.

Consider the ways an ecologist might approach studying the life of honeybees:The behavioural relationship between individuals of a species is behavioural ecology — for example, the study of the queen bee, and how she relates to the w orker beesand the drones. The organized activity of a species is community ecology; for example, the activity of bees assures the pollination of flowering plants. Bee hives additionally produce honey which is consumed by still other species, such as bears.

The relationship between the environment and a species is environmental ecology — for example, the consequences of environmental change on bee activity. Bees may die out due to environmental changes. The environment simultaneously affects and is a consequence of this activity and is thus intertwined with the survival of the species.

DISCIPLINES OF ECOLOGY

Ecology is a broad discipline comprised of many sub-disciplines. A common, broad classification, moving from lowest to highest complexity, where complexity is defined as the number of entities and processes in the system under study, is:

Ecophysiology and Behavioural ecology examine adaptations of the individual to its environment.

Autecology studies the dynamics of populations of a single species.

Community ecology (or synecology) focuses on the interactions between species within an ecological community.

Ecosystem ecology studies the flows of energy and matter through the biotic and abiotic components of ecosystems.

Landscape ecology examines processes and relationship across multiple ecosystems or very large geographic areas.

Ecology can also be sub-divided according to the species of interest into fields such as animal ecology, plant ecology, insect ecology, and so on. Another frequent method of subdivision is by biome studied, e.g., Arctic ecology (or polar ecology), tropical ecology, desert ecology, etc. The primary technique used for investigation is often used to subdivide the discipline into groups such as chemical ecology, genetic ecology, field ecology, statistical ecology, theoretical ecology, and so forth. Note that these different systems are unrelated and often applied at the same time; one could be a theoretical plant community ecologist, or a polar ecologist interested in animal genetics.

FUNDAMENTAL PRINCIPLES OF ECOLOGY

For modern ecologists, ecology can be studied at several levels: population level (individuals of the same species in the same or similar environment), biocenosis level (or community of species), ecosystem level, and biosphere level.

The outer layer of the planet Earth can be divided into several compartments: the hydrosphere (or sphere of water), the lithosphere (or sphere of soils and rocks), and the atmosphere (or sphere of the air). The biosphere (or sphere of life), sometimes described as "the fourth envelope", is all living matter on the planet or that portion of the planet occupied by life. It reaches well into the other three spheres, although there are no permanent inhabitants of the atmosphere. Relative to the volume of the Earth, the biosphere is only the very thin surface layer which extends from 11,000 metres below sea level to 15,000 metres above.

It is thought that life first developed in the hydrosphere, at shallow depths, in the photic zone. Multicellular organisms then appeared and colonized benthic zones. Photosynthetic organisms gradually produced the chemically unstable oxygen-rich atmosphere that characterizes our planet. Terrestrial life developed later, after the ozone layer protecting living beings from UV rays formed. Diversification of terrestrial species is thought to be increased by the continents drifting apart, or alternately, colliding. Biodiversity is expressed at the ecological level (ecosystem), population level (intraspecific diversity), species level (specific diversity), and genetic level. Recently technology has allowed the discovery of the deep ocean vent communities. This remarkable ecological system is not dependent on sunlight but bacteria, utilising the chemistry of the hot volcanic vents, are at the base of its food chain.

The biosphere contains great quantities of elements such as carbon, nitrogen and oxygen. Other elements, such as phosphorus, calcium, and potassium, are also essential to life, yet are present in smaller amounts. At the ecosystem and biosphere levels, there is a continual recycling of all these elements, which alternate between the mineral and organic states.

While there is a slight input of geothermal energy, the bulk of the functioning of the ecosystem is based on the input of solar energy. Plants and photosynthetic microorganisms convert light into chemical energy by the process of photosynthesis, which creates glucose (a simple sugar) and releases free oxygen. Glucose thus becomes the secondary energy source which drives the ecosystem. Some of this glucose is used directly by other organisms for energy. Other sugar molecules can be converted to other molecules such as amino acids. Plants use some of this sugar, concentrated in nectar to entice pollinators to aid them in reproduction.

Cellular respiration is the process by which organisms (like mammals) break the glucose back down into its constituents, water and carbon dioxide, thus regaining the stored energy the sun originally gave to the plants. The proportion of photosynthetic activity of plants and other photosynthesizers to the respiration of other organisms determines the specific composition of the Earth's atmosphere, particularly its oxygen level. Global air currents mix the atmosphere and maintain nearly the same balance of elements in areas of intense biological activity and areas of slight biological activity.

Water is also exchanged between the hydrosphere, lithosphere, atmosphere and biosphere in regular cycles. The oceans are large tanks, which store water, ensure thermal and climatic stability, as well as the transport of chemical elements thanks to large oceanic currents.

For a better understanding of how the biosphere works, and various dysfunctions related to human activity, American scientists simulated the biosphere in a small-scale model, called Biosphere II.

BIOSPHERE

The term "biosphere" was coined by geologist Eduard Suess in 1875, which he defined as:

"The place on earth's surface where life dwells".

While this concept has a geological origin, it is an indication of the impact of both Darwin and Maury on the earth sciences. The biosphere's ecological context comes from the 1920s, preceding the 1935 introduction of the term "ecosystem" by Sir Arthur Tansley (see ecology history). Vernadsky defined ecology as the science of the biosphere. It is an interdisciplinary concept for integrating astronomy, geophysics, meteorology, biogeography, evolution, geology, geochemistry, hydrology and, generally speaking, all life and earth sciences.

A familiar scene on Earth which simultaneously shows the lithosphere, hydrosphere and atmosphere.

Some life scientists and earth scientists use biosphere in a more limited sense. For example, geochemists define the biosphere as being the total sum of living organisms (the "biomass" or "biota" as referred to by biologists and ecologists). In this sense, the biosphere is but one of four separate components of the geochemical model, the other three being lithosphere, hydrosphere, and atmosphere. The narrow meaning used by geochemists is one of the consequences of specialization in modern science. Some might prefer the word ecosphere, coined in the 1960s, as all encompassing of both biological and physical components of the planet.

The Second International Conference on Closed Life Systems defined biospherics as the science and technology of analogs and models of Earth's biosphere; i.e., artificial Earth-like biospheres. Others may include the creation of artificial non-Earth biospheres — for example, human-centered biospheres or a native Martian biosphere — in the field of biospherics.

The biosphere is the life zone of the Earth and includes all living organisms, including man, and all organic matter that has not yet decomposed. Life evolved on earth during its early history between 4.5 and 3.8 billion years ago and the biosphere readily distinguishes our planet from all others in the solar system. The chemical reactions of life (e.g., photosynthesis-respiration, carbonate precipitation, etc.) have also imparted a strong signal on the chemical composition of the atmosphere, transforming the atmosphere from reducing conditions to and oxidizing environment with free oxygen. The biosphere is structured into a hierarchy known as the food chain whereby all life is dependent upon the first tier (i.e. mainly the primary producers that are capable of photosynthesis). Energy and mass is transferred from one level of the food chain to the next with an efficiency of about 10%. All organisms are intrinsically linked to their physical environment and the relationship between an organism and its environment is the study of ecology.

The biosphere can be divided into distinct ecosystems that represent the interactions between a group of organisms forming a trophic pyramid and the environment or habitat in which they live. The biosphere is the outermost part of the planet's shell – including air, land, surface rocks and water – within which life occurs, and which biotic processes in turn alter or transform. From the broadest geophysiological point of view, the biosphere is the global ecological system integrating all living beings and their relationships, including their interaction with the elements of the lithosphere (rocks), hydrosphere (water), and atmosphere (air). This biosphere is postulated to have evolved, beginning through a process of biogenesis or biopoesis, at least some 3.5 billion years ago.

Biomass accounts for about 3.7 kg carbon per square metre of the earth's surface averaged over land and sea, making a total of about 1900 gigatonnes of carbon.

GAIA HYPOTHESIS IN ECOLOGY

After much initial criticism, a modified Gaia hypothesis is now considered within ecological science basically consistent with the planet earth being the ultimate object of ecological study. Ecologists generally consider the biosphere as an ecosystem and the Gaia hypothesis, though a simplification of that original proposed, to be consistent with a modern vision of global ecology, relaying the concepts of biosphere and biodiversity. The Gaia hypothesis has been called geophysiology or Earth system science, which takes into account the interactions between biota, the oceans, the geosphere, and the atmosphere. To promote research and discussion in these fields an organisation, "Gaia Society for Research and Education in Earth System Science" was started.

An example of the change in acceptability of Gaia theories is the Amsterdam declaration of the scientific communities of four international global change research programmes - the International Geosphere-Biosphere Programme (IGBP), the International Human Dimensions Programme on Global Environmental Change (IHDP), the World Climate Research Programme (WCRP) and the international biodiversity programme DIVERSITAS - recognise that, in addition to the threat of significant climate change, there is growing concern over the ever-increasing human modification of other aspects of the global environment and the consequent implications for human well-being.

They State

"Research carried out over the past decade under the auspices of the four programmes to address these concerns has shown that:

The Earth System behaves as a single, self-regulating system comprised of physical, chemical, biological and human components. The interactions and feedbacks between the component parts are complex and exhibit multi-scale temporal and spatial variability. The understanding of the natural dynamics of the Earth System has advanced greatly in recent years and provides a sound basis for evaluating the effects and consequences of human-driven change.

Human activities are significantly influencing Earth's environment in many ways in addition to greenhouse gas emissions and climate change. Anthropogenic changes to Earth's land surface, oceans, coasts and atmosphere and to biological diversity, the water cycle and biogeochemical cycles are clearly identifiable beyond natural variability. They are equal to some of the great forces of nature in their extent and impact. Many are accelerating. Global change is real and is happening now.

Global change cannot be understood in terms of a simple cause-effect paradigm. Human-driven changes cause multiple effects that cascade through the Earth System in complex ways. These effects interact with each other and with local- and regional-scale changes in multidimensional patterns that are difficult to understand and even more difficult to predict. Surprises abound.

Earth System dynamics are characterised by critical thresholds and abrupt changes. Human activities could inadvertently trigger such changes with severe consequences for Earth's environment and inhabitants. The Earth System has operated in different states over the last half million years, with abrupt transitions (a decade or less) sometimes occurring between them. Human activities have the potential to switch the Earth System to alternative modes of operation that may prove irreversible and less hospitable to humans and other life. The probability of a human-driven abrupt change in Earth's environment has yet to be quantified but is not negligible.

In terms of some key environmental parameters, the Earth System has moved well outside the range of the natural variability exhibited over the last half million years at least. The nature of changes now occurring simultaneously in the Earth System, their magnitudes and rates of change are unprecedented. The Earth is currently operating in a no-analogue state."

Sir Crispin Tickell in the the 46th Annual Bennett Lecture for the 50th Anniversary of Geology at the University of Leicester in his recent talk "Earth Systems Science: Are We Pushing Gaia Too Hard?" stated "as a theory, Gaia is now winning."

He continued "The same goes for the earth systems science which is now the concern of the Geological Society of London (with which the Gaia Society recently merged). Whatever the label, earth systems science, or Gaia, has now become a major subject of enquiry and research, and no longer has to justify itself."

These findings would seem to be fully in accord with the Gaia theory. Despite this endorsement, the late Bill Hamilton, one of the founders of modern Darwinism, whilst conceding the empirical basis of the planetary homeostatic processes on which Gaia is based, states that it is a theory still awaiting its Copernicus.

BIODIVERSITY

Biodiversity is the variety and differences among living organisms from all sources, including terrestrial, marine, and other aquatic ecosystems and the ecological complexes of which they are a part. This includes genetic diversity within and between species and of ecosystems. Thus, in essence, biodiversity represents all life. India is one of the mega biodiversity centres in the world and has two of the world's 18 'biodiversity hotspots' located in the Western Ghats and in the Eastern Himalayas. The forest cover in these areas is very dense and diverse and of pristine beauty, and incredible biodiversity.

Biodiversity is the variety of all living things; the different plants, animals and micro organisms, the genetic information they contain and the ecosystems they form.

Biodiversity is usually explored at three levels - genetic diversity, species diversity and ecosystem diversity. These three levels work together to create the complexity of life on Earth.

Biodiversity or biological diversity is the variation of taxonomic life forms within a given ecosystem, biome or for the entire Earth. Biodiversity is often a measure of the health of biological systems to indicate the degree to which the aggregate of historical species are viable versus extinct.

Fig. Biodiversity

EVOLUTION AND MEANING OF THE TERM

Biodiversity is a neologism and a portmanteau word, from bio and diversity.The Science Division of The Nature Conservancy used the term "natural diversity" in a 1974 study, "The Preservation of Natural Diversity." The term biological diversity was used even before that by conservation scientists like Robert E. Jenkins and Thomas Lovejoy. The word biodiversity itself may have been coined by W.G. Rosen in 1985 while planning the National Forum on Biological Diversity organized by the National Research Council (NRC) which was to be held in 1986, and first appeared in a publication in 1988 when entomologist E. O. Wilson used it as the title of the proceedings of that forum. The word biodiversity was deemed more effective in terms of communication than biological diversity.

Since 1986 the terms and the concept have achieved widespread use among biologists, environmentalists, political leaders, and concerned citizens worldwide. It is generally used to equate to a concern for the natural environment and nature conservation. This use has coincided with the expansion of concern over extinction observed in the last decades of the 20th century.

The term "natural heritage" predates "biodiversity", though it is a less scientific term and more easily comprehended in some ways by the wider audience interested in conservation. "Natural Heritage" was used when Jimmy Carter set up the Georgia Heritage Trust while he was governor of Georgia; Carter's trust dealt with both natural and cultural heritage. It would appear that Carter picked the term up from Lyndon Johnson, who used it in a 1966 Message to Congress. "Natural Heritage" was picked up by the Science Division of The Nature Conservancy when, under Jenkins, it launched in 1974 the network of State Natural Heritage Programmes. When this network was extended outside the USA, the term "Conservation Data Centre" was suggested by Guillermo Mann and came to be preferred.

Definitions

The most straightforward definition is "variation of life at all levels of biological organization". A second definition holds that biodiversity is a measure of the relative diversity among organisms present in different ecosystems. "Diversity" in this definition includes diversity within a species and among species, and comparative diversity among ecosystems.

A third definition that is often used by ecologists is the "totality of genes, species, and ecosystems of a region". An advantage of this definition is that it seems to describe most circumstances and present a unified view of the traditional three levels at which biodiversity has been identified:

Genetic Diversity: Diversity of genes within a species. There is a genetic variability among the populations and the individuals of the same species.

Species Diversity: diversity among species in an ecosystem. "Biodiversity hotspots" are excellent examples of species diversity.

Ecosystem Diversity: Diversity at a higher level of organization, the ecosystem. To do with the variety of ecosystems on Earth.

This third definition, which conforms to the traditional five organization layers in biology, provides additional justification for multilevel approaches.

The 1992 United Nations Earth Summit in Rio de Janeiro defined "biodiversity" as "the variability among living organisms from all sources, including, 'inter alia', terrestrial, marine, and other aquatic ecosystems, and the ecological complexes of which they are part: this includes diversity within species, between species and of ecosystems". This is, in fact, the closest thing to a single legally accepted definition of biodiversity, since it is the definition adopted by the United Nations Convention on Biological Diversity. The parties to this convention include all the countries on Earth, with the exception of Andorra, Brunei Darussalam, the Holy See, Iraq, Somalia, and the United States of America.

If the gene is the fundamental unit of natural selection, according to E. O. Wilson, the real biodiversity is genetic diversity. For geneticists, biodiversity is the diversity of genes and organisms. They study processes such as mutations, gene exchanges, and genome dynamics that occur at the DNA level and generate evolution.

For biologists, biodiversity is the gamut of organisms and species and their interactions. Organisms appear and become extinct; sites are colonized and some species develop social organizations to improve their varied strategies of reproduction.

For ecologists, biodiversity is also the diversity of durable interactions among species. It not only applies to species, but also to their immediate environment (biotope) and their larger ecoregion. In each ecosystem, living organisms are part of a whole, interacting with not only other organisms, but also with the air, water, and soil that surround them.

MEASUREMENT OF BIODIVERSITY

Biodiversity is a broad concept, so a variety of objective measures have been created in order to empirically measure biodiversity. Each measure of biodiversity relates to a particular use of the data.

For practical conservationists, this measure should quantify a value that is broadly shared among locally affected people. For others, a more economically defensible definition should allow the ensuring of continued possibilities for both adaptation and future use by people, assuring environmental sustainability.

As a consequence, biologists argue that this measure is likely to be associated with the variety of genes. Since it cannot always be said which genes are more likely to prove beneficial, the best choice for conservation is to assure the persistence of as many genes as possible. For ecologists, this latter approach is sometimes considered too restrictive, as it prohibits ecological succession.

Biodiversity is usually plotted as taxonomic richness of a geographic area, with some reference to a temporal scale. Whittaker described three common metrics used to measure species-level biodiversity, encompassing attention to species richness or species evenness:

Species richness - the most primitive of the indices available.

Simpson index

Shannon index

There are three other indices which are used by ecologists:

Alpha diversity refers to diversity within a particular area, community or ecosystem, and is measured by counting the number of taxa within the ecosystem (usually species)

Beta diversity is species diversity between ecosystems; this involves comparing the number of taxa that are unique to each of the ecosystems.

Gamma diversity is a measure of the overall diversity for different ecosystems within a region.

DISTRIBUTION OF BIODIVERSITY

Biodiversity is not distributed evenly on Earth. It is consistently richer in the tropics and in other localized regions such as the California Floristic Province. As one approaches polar regions one generally finds fewer species. Flora and fauna diversity depends on climate, altitude, soils and the presence of other species. In the year 2006 large numbers of the Earth's species are formally classified as rare or endangered or threatened species; moreover, most scientists estimate that there are millions more species actually endangered which have not yet been formally recognized.

A biodiversity hotspot is a region with a high level of endemic species. These biodiversity hotspots were first identified by Dr. Norman Myers in two articles in the scientific journal The Environmentalist. Hotspots unfortunately tend to occur near areas of dense human habitation, leading to threats to their many endemic species. As a result of the pressures of the rapidly growing human population, human activity in many of these areas is increasing dramatically. Most of these hotspots are located in the tropics and most of them are forests.

For example, Brazil's Atlantic Forest contains roughly 20,000 plant species, 1350 vertebrates, and millions of insects, about half of which occur nowhere else in the world. The Madagascar dry deciduous forests and lowland rainforests possess a very high ratio of species endemism and biodiversity, arising from the fact that this island separated from mainland Africa 65 million years ago.

Many regions of high biodiversity (as well as high endemism) arise from very specialized habitats which require unusual adaptation mechanisms. For example the peat bogs of Northern Europe and the alvar regions such as the Stora Alvaret on Oland, Sweden host a large diversity of plants and animals, many of whom are not found elsewhere.

BIODIVERSITY AND EVOLUTION

Biodiversity found on Earth today is the result of 4 billion years of evolution. The origin of life is not well known to science, though limited evidence suggests that life may already have been well-established only a few 100 million years after the formation of the Earth. Until approximately 600 million years ago, all life consisted of bacteria and similar single-celled organisms.

The history of biodiversity during the Phanerozoic (the last 540 million years), starts with rapid growth during the Cambrian explosion—a period during which nearly every phylum of multicellular organisms first appeared. Over the next 400 million years or so, global diversity showed little overall trend, but was marked by periodic, massive losses of diversity classified as mass extinction events.

The apparent biodiversity shown in the fossil record suggests that the last few million years include the period of greatest biodiversity in the Earth's history. However, not all scientists support this view, since there is considerable uncertainty as to how strongly the fossil record is biased by the greater availability and preservation of recent geologic sections. Some (e.g. Alroy et al. 2001) argue that corrected for sampling artifacts, modern biodiversity is not much different from biodiversity 300 million years ago. Estimates of the present global macroscopic species diversity vary from 2 million to 100 million species, with a best estimate of somewhere near 10 million.

Most biologists agree however that the period since the emergence of humans is part of a new mass extinction, the Holocene extinction event, caused primarily by the impact humans are having on the environment. At present, the number of species estimated to have gone extinct as a result of human action is still far smaller than are observed during the major mass extinctions of the geological past. However, it has been argued that the present rate of extinction is sufficient to create a major mass extinction in less than 100 years. Others dispute this and suggest that the present rate of extinctions could be sustained for many thousands of years before the loss of biodiversity matches the more than 20% losses seen in past global extinction events.

New species are regularly discovered (on average about three new species of birds each year) and many, though discovered, are not yet classified (an estimate states that about 40% of freshwater fish from South America are not yet classified). Most of the terrestrial diversity is found in tropical forests.

BENEFITS OF BIODIVERSITY

There are a multitude of benefits of biodiversity in the sense of one diverse group aiding another such as:

Food and Drink

Biodiversity provides food for humans. About 80 percent of our food supply comes from just 20 kinds of plant. Although many kinds of animal are utilised as food, again most consumption is focused on a few species.

There is vast untapped potential for increasing the range of food products suitable for human consumption.

Medicines

A significant proportion of drugs are derived, directly or indirectly, from biological sources; in most cases these medicines can not presently be synthesized in a laboratory setting. Moreover, only a small proportion of the total diversity of plants has been thoroughly investigated for potential sources of new drugs. Many Medicines and antibiotics are also derived from microorganisms.

Industrial Materials

A wide range of industrial materials are derived directly from biological resources. These include building materials, fibres, dyes, resins, gums, adhesives, rubber and oil. There is enormous potential for further research into sustainably utilising materials from a wider diversity of organisms.

Other Ecological Services

Biodiversity provides many ecosystem services that are often not readily visible. It plays a part in regulating the chemistry of our atmosphere and water supply. Biodiversity is directly involved in recycling nutrients and providing fertile soils. Experiments with controlled environments have shown that humans cannot easily build ecosystems to support human needs; for example insect pollination cannot be mimicked by man-made construction, and that activity alone represents tens of billions of dollars in ecosystem services per annum to mankind.

Leisure, Cultural and Aesthetic Value

Many people derive value from biodiversity through leisure activities such as enjoying a walk in the countryside, birdwatching or natural history programmes on television.

Biodiversity has inspired musicians, painters, sculptors, writers and other artists. Many cultural groups view themselves as an integral part of the natural world and show respect for other living organisms.

THREATS TO BIODIVERSITY

During the last century, erosion of biodiversity has been increasingly observed. Some studies show that about one of eight known plant species is threatened with extinction. Some estimates put the loss at up to 140,000 species per year (based on Species-area theory) and subject to discussion. Most of the species extinctions from 1000 AD to 2000 AD are due to human activities, in particular destruction of plant and animal habitats. Almost all scientists acknowledge that the rate of species loss is greater now than at any time in human history, with extinctions occurring at rates hundreds of times higher than background extinction rates.

Elevated rates of extinction are being driven by human consumption of organic resources, especially related to tropical forest destruction. While most of the species that are becoming extinct are not food species, their biomass is converted into human food when their habitat is transformed into pasture, cropland, and orchards. It is estimated that more than 40% of the Earth's biomass is tied up in only the few species that represent humans, our livestock and crops. Because an ecosystem decreases in stability as its species are made extinct, these studies warn that the global ecosystem is destined for collapse if it is further reduced in complexity. Factors contributing to loss of biodiversity are: overpopulation, deforestation, pollution (air pollution, water pollution, soil contamination) and global warming or climate change, driven by human activity. These factors, while all stemming from overpopulation, produce a cumulative impact upon biodiversity.

Some characterize loss of biodiversity not as ecosystem degradation but by conversion to trivial standardized ecosystems (e.g., monoculture following deforestation). In some countries lack of property rights or access regulation to biotic resources necessarily leads to biodiversity loss (degradation costs having to be supported by the community).

The widespread introduction of exotic species by humans is a potent threat to biodiversity. When exotic species are introduced to ecosystems and establish self-sustaining populations, the endemic species in that ecosystem, that have not evolved to cope with the exotic species, may not survive. The exotic organisms may be either predators, parasites, or simply aggressive species that deprive indigenous species of nutrients, water and light. These exotic or invasive species often have features due to their evolutionary background and environment that makes them very competitive, and similarly makes endemic species very defenceless and/or uncompetitive against these exotic species.

The rich diversity of unique species across many parts of the world exist only because they are separated by barriers, particularly seas and oceans, from other species of other land masses, particularly the highly fecund, ultra-competitive, generalist "super-species". These are barriers that could never be crossed by natural processes, except for many millions of years in the future through continental drift. Howeve: humans have invented ships and aeroplanes, and now have the power to bring into contact species that never have met in their evolutionary history, and on a time scale of days, unlike the centuries that historically have accompanied major animal migrations.

As a consequence of the above, if humans continue to combine species from different ecoregions, there is the potential that the world's ecosystems will end up dominated by a very few, aggressive, cosmopolitan "super-species".

In 2004, an international team of scientists estimated that 15 percent to 37 percent of species would become extinct by 2050 because of global warming. "We need to limit climate change or we wind up with a lot of species in trouble, possibly extinct," said Dr. Lee Hannah, a co-author of the paper and chief climate change biologist at the Centre for Applied Biodiversity Science at Conservation International.

BIODIVERSITY MANAGEMENT: CONSERVATION, PRESERVATION AND PROTECTION

The conservation of biological diversity has become a global concern. Although not everybody agrees on extent and significance of current extinction, most consider biodiversity essential. There are basically two main types of conservation options, in-situ conservation and ex-situ conservation. In-situ is usually seen as the ideal conservation strategy. However, its implementation is sometimes unfeasible. For example, destruction of rare or endangered species' habitats sometimes requires ex-situ conservation efforts. Furthermore, ex-situ conservation can provide a backup solution to in-situ conservation projects. Some believe both types of conservation are required to ensure proper preservation. An example of an in-situ conservation effort is the setting-up of protection areas. Examples of ex-situ conservation efforts, by contrast, would be planting germplasts in seedbanks, or growing the Wollemi Pine in nurseries. Such efforts allow the preservation of large populations of plants with minimal genetic erosion.

At national levels a Biodiversity Action Plan is sometimes prepared to state the protocols necessary to protect an individual species. Usually this plan also details extant data on the species and its habitat. In the USA such a plan is called a Recovery Plan.

The threat to biological diversity was among the hot topics discussed at the UN World Summit for Sustainable Development, in hope of seeing the foundation of a Global Conservation Trust to help maintain plant collections.

JURIDICAL STATUS OF BIOLOGICAL DIVERSITY

Biodiversity must be evaluated and its evolution analysed (through observations, inventories, conservation...) then it must be taken into account in political decisions. It is beginning to receive a juridical setting.

"Law and ecosystems" relationship is very ancient and has consequences for biodiversity. It is related to property rights, private and public. It can define protection for threatened ecosystems, but also some rights and duties (for example, fishing rights, hunting rights).

"Laws and species" is a more recent issue. It defines species that must be protected because threatened by extinction. Some people question application of these laws. The U.S. Endangered Species Act is an example of an attempt to address the "law and species" issue.

"Laws and genes" is only about a century old. While the genetic approach is not new (domestication, plant traditional selection methods), progress made in the genetic field in the past 20 years lead to the obligation to tighten laws. With the new technologies of genetic and genetic engineering, people are going through gene patenting, processes patenting, and a totally new concept of genetic resource. A very hot debate today seeks to define whether the resource is the gene, the organism, the DNA or the processes.

The 1972 UNESCO convention established that biological resources, such as plants, were the common heritage of mankind. These rules probably inspired the creation of great public banks of genetic resources, located outside the source-countries.

New global agreements (e.g.Convention on Biological Diversity), now give sovereign national rights over biological resources (not property). The idea of static conservation of biodiversity is disappearing and being replaced by the idea of dynamic conservation, through the notion of resource and innovation.

The new agreements commit countries to conserve biodiversity, develop resources for sustainability and share the benefits resulting from their use. Under new rules, it is expected that bioprospecting or collection of natural products has to be allowed by the biodiversity-rich country, in exchange for a share of the benefits.

Sovereignty principles can rely upon what is better known as Access and Benefit Sharing Agreements (ABAs). The

Convention on Biodiversity spirit implies a prior informed consent between the source country and the collector, to establish which resource will be used and for what, and to settle on a fair agreement on benefit sharing. Bioprospecting can become a type of biopiracy when those principles are not respected.

Uniform approval for use of biodiversity as a legal standard has not been achieved, however. At least one legal commentator has argued that biodiversity should not be used as a legal standard, arguing that the multiple layers of scientific uncertainty inherent in the concept of biodiversity will cause administrative waste and increase litigation without promoting preservation goals.

THE FOUNDER EFFECT

The field of biodiversity research has often been criticized for being overly defined by the personal interests of the founders (i.e. terrestrial mammals) giving a narrow focus, rather than extending to other areas where it could be useful. This is termed the founder effect by Norse and Irish, (1996). (This was a play on words: the founder effect in ecology typically refers to the genetic outcome when a small population establishes an isolated breeding group). France and Rigg reviewed the biodiversity literature in 1998 and found that there was a significant lack of papers studying marine ecosystems, leading them to dub marine biodiversity research the sleeping hydra. More work has been carried out for accessible, diverse coastal systems such as coral reefs than for inaccessible, species-poor deep sea areas.

It has been easier to mobilise public opinion and national legislation for the terrestrial realm, which has higher visibility and falls within countries' territorial boundaries. Marine conservation involves having to pioneer new and international mechanisms of protection as well as solving methodological problems in marine biology relating to marine ecosystem classification and data-gathering on some of the earth's most difficult species to access and monitor.

SIZE BIAS

Biodiversity researcher Sean Nee points out that the vast majority of Earth's biodiversity is microbial, and that contemporary biodiversity physics is "firmly fixated on the visible world" (Nee uses "visible" as a synonym for macroscopic). For example, microbial life is very much more metabolically and environmentally diverse than multicellular life (see extremophile). Nee has stated: "On the tree of life, based on analyses of small-subunit ribosomal RNA, visible life consists of barely noticeable twigs. This should not be surprising — invisible life had at least three billion years to diversify and explore evolutionary space before the 'visibles' arrived".

The reply to this, however, is that biodiversity conservation has never focused exclusively on visible (in this sense) species. From the very beginning, the classification and conservation of natural communities or ecosystem types has been a central part of the effort. The thought behind this has been that since invisible (in this sense) diversity is, due to lack of taxonomy, impossible to treat in the same manner as visible diversity, the best that can be done is to preserve a diversity of ecosystem types, thereby preserving as well as possible the diversity of invisible organisms.

Genetic diversity is the variety of genes within a species. Each species is made up of individuals that have their own particular genetic composition. This means a species may have different populations, each having different genetic compositions. To conserve genetic diversity, different populations of a species must be conserved.

Genes are the basic units of all life on Earth. They are responsible for both the similarities and the differences between organisms.

Not all groups of animals have the same degree of genetic diversity. Kangaroos, for example, come from recent evolutionary lines and are genetically very similar. Carnivorous marsupials, called dasyurids, come from more ancient lines and are genetically far more diverse. Some scientists believe that we should concentrate on saving more genetically diverse groups, such as dasyurids, which include the Tasmanian Devil, the Numbat and quolls.

Species diversity is the variety of species within a habitat or a region. Some habitats, such as rainforests and coral reefs, have many species. Others, such as salt flats or a polluted stream, have fewer.

Species are grouped together into families according to shared characteristics. Invertebrates - animals without backbones - make up about 99% of all animal species, and most of these are insects. Invertebrates include crabs, snails, worms, corals and seastars, as well as insects, such as beetles and flies. Insects fill many vital roles in ecosystems as pollinators, recyclers of nutrients, scavengers and food for others.

Ecosystem diversity is the variety of ecosystems in a given place. An ecosystem is a community of organisms and their physical environment interacting together.An ecosystem can cover a large area, such as a whole forest, or a small area, such as a pond.

An ecosystem is a community of organisms and their physical environment interacting together. An ecosystem may be as large as the Great Barrier Reef or as small as the back of a spider crab's shell, which provides a home for plants and other animals, such as sponges, algae and worms.

DYNAMICS AND STABILITY

Ecological factors which affect dynamic change in a population or species in a given ecology or environment are usually divided into two groups: abiotic and biotic.

Abiotic factors are geological, geographical, hydrological and climatological parameters. A biotope is an environmentally uniform region characterized by a particular set of abiotic ecological factors. Specific abiotic factors include:

- Water, which is at the same time an essential element to life and a milieu

- Air, which provides oxygen, nitrogen, and carbon dioxide to living species and allows the dissemination of pollen and spores
- Soil, at the same time source of nutriment and physical support
- Soil pH, salinity, nitrogen and phosphorus content, ability to retain water, and density are all influential
- Temperature, which should not exceed certain extremes, even if tolerance to heat is significant for some species
- Light, which provides energy to the ecosystem through photosynthesis
- Natural disasters can also be considered abiotic

Biocenose, or community, is a group of populations of plants, animals, micro-organisms. Each population is the result of procreations between individuals of same species and cohabitation in a given place and for a given time. When a population consists of an insufficient number of individuals, that population is threatened with extinction; the extinction of a species can approach when all biocenoses composed of individuals of the species are in decline. In small populations, consanguinity (inbreeding) can result in reduced genetic diversity that can further weaken the biocenose.

Biotic ecological factors also influence biocenose viability; these factors are considered as either intraspecific and interspecific relations.

Intraspecific relations are those which are established between individuals of the same species, forming a population. They are relations of co-operation or competition, with division of the territory, and sometimes organization in hierarchical societies.

An ant lion lies in wait under its pit trap, built in dry dust under a building, awaiting unwary insects that falls in. Many pest insects are partly or wholly controlled by other insect predators.

Interspecific relations—interactions between different species—are numerous, and usually described according to their beneficial, detrimental or neutral effect (for example, mutualism (relation ++) or competition (relation --). The most significant relation is the relation of predation (to eat or to be eaten), which leads to the essential concepts in ecology of food chains (for example, the grass is consumed by the herbivore, itself consumed by a carnivore, itself consumed by a carnivore of larger size).

A high predator to prey ratio can have a negative influence on both the predator and prey biocenoses in that low availability of food and high death rate prior to sexual maturity can decrease (or prevent the increase of) populations of each, respectively. Selective hunting of species by humans which leads to population decline is one example of a high predator to prey ratio in action. Other interspecific relations include parasitism, infectious disease and competition for limiting resources, which can occur when two species share the same ecological niche.

The existing interactions between the various living beings go along with a permanent mixing of mineral and organic substances, absorbed by organisms for their growth, their maintenance and their reproduction, to be finally rejected as waste. These permanent recyclings of the elements (in particular carbon, oxygen and nitrogen) as well as the water are called biogeochemical cycles.

They guarantee a durable stability of the biosphere (at least when unchecked human influence and extreme weather or geological phenomena are left aside). This self-regulation, supported by negative feedback controls, ensures the perenniality of the ecosystems. It is shown by the very stable concentrations of most elements of each compartment. This is referred to as homeostasis. The ecosystem also tends to evolve to a state of ideal balance, reached after a succession of events, the climax (for example a pond can become a peat bog).

SPATIAL RELATIONSHIPS AND SUBDIVISIONS OF LAND

Ecosystems are not isolated from each other, but are interrelated. For example, water may circulate between ecosystems by the means of a river or ocean current. Water itself, as a liquid medium, even defines ecosystems. Some species, such as salmon or freshwater eels move between marine systems and fresh-water systems. These relationships between the ecosystems lead to the concept of a biome.

A biome is a homogeneous ecological formation that exists over a large region as tundra or steppes. The biosphere comprises all of the Earth's biomes — the entirety of places where life is possible — from the highest mountains to the depths of the oceans.

Biomes correspond rather well to subdivisions distributed along the latitudes, from the equator towards the poles, with differences based on to the physical environment (for example, oceans or mountain ranges) and to the climate. Their variation is generally related to the distribution of species according to their ability to tolerate temperature and/or dryness. For example, one may find photosynthetic algae only in the photic part of the ocean (where light penetrates), while conifers are mostly found in mountains. Though this is a simplification of more complicated scheme, latitude and altitude approximate a good representation of the distribution of biodiversity within the biosphere. Very generally, the richness of biodiversity (as well for animal than plant species) is decreasing most rapidly near the equator and less rapidly as one approaches the poles.

The biosphere may also be divided into ecozone, which are very well defined today and primarily follow the continental borders. The ecozones are themselves divided into ecoregions, though there is not agreement on their limits.

ECOSYSTEM PRODUCTIVITY

In an ecosystem, the connections between species are generally related to food and their role in the food chain. There are three categories of organisms:

Producers: usually plants which are capable of photosynthesis but could be other organisms such as bacteria around ocean vents that are capable of chemosynthesis.

Consumers: animals, which can be primary consumers (herbivorous), or secondary or tertiary consumers (carnivorous).

Decomposers: bacteria, mushrooms which degrade organic matter of all categories, and restore minerals to the environment.

These relations form sequences, in which each individual consumes the preceding one and is consumed by the one following, in what are called food chains or food network. In a food network, there will be fewer organisms at each level as one follows the links of the network up the chain.

These concepts lead to the idea of biomass (the total living matter in a given place), of primary productivity (the increase in the mass of plants during a given time) and of secondary productivity (the living matter produced by consumers and the decomposers in a given time).

These two last ideas are key, since they make it possible to evaluate the load capacity – the number of organisms which can be supported by a given ecosystem. In any food network, the energy contained in the level of the producers is not completely transferred to the consumers. And the higher one goes up the chain, the more energy and resources is lost and consumed. Thus, from an energy—and environmental—point of view, it is more efficient for humans to be primary consumers (to subsist from vegetables, grains, legumes, fruit, cotton, etc.) than as secondary consumers (from eating herbivores, omnivores, or their products, such as milk, chickens, cattle, sheep, etc.) and still more so than as a tertiary consumer (from consuming carnivores, omnivores, or their products, such as fur, pigs, snakes, alligators, etc.). An ecosystem(s) is unstable when the load capacity is overrun and is especially unstable when a population doesn't have an ecological niche and overconsumers.

The productivity of ecosystems is sometimes estimated by comparing three types of land-based ecosystems and the total of aquatic ecosystems:

The forests (1/3 of the Earth's land area) contain dense biomasses and are very productive. The total production of the world's forests corresponds to half of the primary production.

Savannas, meadows, and marshes (1/3 of the Earth's land area) contain less dense biomasses, but are productive. These ecosystems represent the major part of what humans depend on for food.

Extreme ecosystems in the areas with more extreme climates – deserts and semi-deserts, tundra, alpine meadows, and steppes – (1/3 of the Earth's land area) have very sparse biomasses and low productivity

Finally, the marine and fresh water ecosystems (3/4 of Earth's surface) contain very sparse biomasses (apart from the coastal zones).

Humanity's actions over the last few centuries have seriously reduced the amount of the Earth covered by forests (deforestation), and have increased agro-ecosystems (agriculture). In recent decades, an increase in the areas occupied by extreme ecosystems has occurred (desertification).

Ecological Crisis

Generally, an ecological crisis occurs with the loss of adaptive capacity when the resilience of an environment or of a species or a population evolves in a way unfavourable to coping with perturbations that interfere with that ecosystem, landscape or species survival.

It may be that the environment quality degrades compared to the species needs, after a change in an abiotic ecological factor (for example, an increase of temperature, less significant rainfalls).

It may be that the environment becomes unfavourable for the survival of a species (or a population) due to an increased pressure of predation (for example overfishing).

Lastly, it may be that the situation becomes unfavourable to the quality of life of the species (or the population) due to a rise in the number of individuals (overpopulation).

Ecological crises may be more or less brutal (occurring within a few months or taking as long as a few million years). They can also be of natural or anthropic origin. They may relate to one unique species or to many species.

Lastly, an ecological crisis may be local (as an oil spill) or global (a rise in the sea level due to global warming).

According to its degree of endemism, a local crisis will have more or less significant consequences, from the death of many individuals to the total extinction of a species. Whatever its origin, disappearance of one or several species often will involve a rupture in the food chain, further impacting the survival of other species.

In the case of a global crisis, the consequences can be much more significant; some extinction events showed the disappearance of more than 90% of existing species at that time. However, it should be noted that the disappearance of certain species, such as the dinosaurs, by freeing an ecological niche, allowed the development and the diversification of the mammals. An ecological crisis thus paradoxically favored biodiversity.

Sometimes, an ecological crisis can be a specific and reversible phenomenon at the ecosystem scale. But more generally, the crises impact will last. Indeed, it rather is a connected series of events, that occur till a final point. From this stage, no return to the previous stable state is possible, and a new stable state will be set up gradually. Lastly, if an ecological crisis can cause extinction, it can also more simply reduce the quality of life of the remaining individuals. Thus, even if the diversity of the human population is sometimes considered threatened (see in particular indigenous people), few people envision human disappearance at short span. However, epidemic diseases, famines, impact on health of reduction of air quality, food crises, reduction of living space, accumulation of toxic or non degradable wastes, threats on keystone species (great apes, panda, whales) are also factors influencing the well-being of people.

During the past decades, this increasing responsibility of humanity in some ecological crises has been clearly observed. Due to the increases in technology and a rapidly increasing population, humans have more influence on their own environment than any other ecosystem engineer.

Some usually quoted examples as ecological crises are:

Permian-Triassic extinction event 250 million of years ago

Cretaceous-Tertiary extinction event 65 million years ago

Global warming related to the Greenhouse effect. Warming could involve flooding of the Asian deltas multiplication of extreme weather phenomena and changes in the nature and quantity of the food resources.

Ozone layer hole issue.

Deforestation and desertification, with disappearance of many species.

The nuclear meltdown at Chernobyl in 1986 caused the death of many people and animals from cancer, and caused mutations in a large number of animals and people. The area around the plant is now abandoned by humans because of the large amount of radiation generated by the meltdown. Twenty years after the accident, the animals have returned.

BIOMS

Most of us are confused when it comes to the words ecosystem and biome. What's the difference? There is a slight difference between the two words. An ecosystem is much smaller than a biome. Conversely, a biome can be thought of many similar ecosystems throughout the world grouped together. An ecosystem can be as large as the Sahara Desert, or as small as a puddle or vernal pool.

Ecosystems are dynamic interactions between plants, animals, and microorganisms and their environment working together as a functional unit. Ecosystems will fail if they do not remain in balance. No community can carry more organisms than its food, water, and shelter can accomodate. Food and territory are often balanced by natural phenomena such as fire, disease, and the number of predators. Each organism has its own niche, or role, to play.

WHAT IS A BIOME?

A biome is a large area with similar flora, fauna, and microorganisms. Most of us are familiar with the tropical rainforests, tundra in the arctic regions, and the evergreen trees in the coniferous forests. Each of these large communities contain species that are adapted to its varying conditions of water, heat, and soil. For instance, polar bears thrive in the arctic while cactus plants have a thick skin to help preserve water in the hot desert.

THE MAJOR BIOMES

- Mountains (High Elevation)
- Temperate Forest
- Desert
- Cold Climate Forest
- Savannah
- Tundra
- Marine/Island
- Tropical Dry Forest
- Grassland
- Tropical Rainforest

Mountains: The High Elevation Biome

Mountains are a common sight on this planet. They make up one-fifth of the world's landscape, and provide homes to at least one-tenth of the world's people. Furthermore, 2 billion people depend on mountain ecosystems for most of their food, hydroelectricity, timber, and minerals. About 80 per cent of our planet's fresh water originates in the mountains. Since about half of the world's people are reliant upon mountains for fresh water, and in this time of increasing water scarcity, it is becoming increasingly important to protect the mountain biome.

All mountain ecosystems have one major characteristic in common—rapid changes in altitude, climate, soil, and vegetation over very short distances. Mountain ecosystems sport a high range of biodiversity, and are also a home to many of our planet's ethnic minorities. These cultures are sometimes 'protected' due to the challenging environment to produce a living, but others are not. More and more these indigenous people are being kicked out of their homes due to population and commercial growth, logging, and mining. An example of the mountain's wide variety of organisms can be seen in California's Sierra Nevada range (the home of Yosemite, which is pictured above, far left). It has been estimated that this range alone houses 10,000 to 15,000 DIFFERENT species of plants and animals! This is all mainly due to elevation changes, which produces belts, or zones, of differing climates, soils, and plant life.

Rainfall varies greatly across the world's montane (mountain) biomes, ranging from very wet to very dry. However in all the biomes comes swift weather changes. For example, in just a few minutes a thunder storm can roll in when the sky was perfectly clear, and in just a few hours the temperatures can drop from extremely hot temperatures to temperatures that are below freezing.

The world's mountains provide a home to several thousand different ethnic groups. The mountain people, which mainly consist of indigenous people, ethnic minorities, and refugees, have been able to cope with this harsh environment of the mountain ecosystem. They live as nomads, hunters, foragers, traders, small farmers, loggers, and miners, etc. Most mountain people all share one attribute – material poverty. However, what they lack in material wealth they make up in community life. They have been able to live off the land without widespread destruction and deforestation. Plant and animal species have been preserved by these people. For instance, in India's Garhwal Himalaya, local women were recently successful in identifying over 145 species of plants that had been destroyed by commercial logging and limestone mining; the national foresters could only list 25! Unfortunately, these cultures have been subjected to discrimination and other violations of human rights. They have been called degrading words such as 'hillbillies' (United States), 'oberwalder' (Austria), 'kohestani' (Afghanistan), and 'bhotias' (India). We need to learn not only how to preserve the biological diversity in the mountains, but the cultural diversity also.

The Himalayan Yew, a slow-growing conifer, is currently on the World Wildlife Fund's list of the ten most endangered animals. This plant can be found throughout Bhutan, Afghanistan, India, Nepal, Burma, and maybe China. Taxol, which is promising to be a drug which can help cure cancer, is present in both the Pacific and Himalayan varieties. Found in the world's highest mountain range, the Himalayan Yew is extremely rare because of heavy deforestation and harvesting for Taxol extraction, without replanting. About 10 kilograms of yew leaves, bark, and needles will only produce one gram of the drug! New controls need to be imposed on this plant to make sure it is replanted and our supply remains sustainable, otherwise a valuable resource, and possible cure for cancer, may be lost.

Commercial industries, especially large mines and hydropower projects, cause exceptional damage in mountains. This is because many companies are ignorant of the fragility of the ecosystems and rights of local communities. The U.N. Food and Agriculture Organization (FAO) discovered that during the past decade, tropical mountain forests have had both the fastest rates of both annual population growth and deforestation. Somehow, we need to come up with a way to find a compromise between preserving the cultural and biological diversity in mountains, and using them as a valuable resource. After all, if the mountains are exploited until they run dry, there will be no more resources for future generations. Currently, only 8 per cent of all mountains are protected in some form. If our world's highest mountains are able to inspire the greatest of mountain climbers to accomplish great feats, we should provide no lesser commitment to preserving the fragile ecosystems and endangered cultures which lie within them.

In Glacier National Park, near Logan's Pass (pictured above, middle), there are many Rocky Mountain goats who don't make their home in their natural ecosystem, but lounge around the parking lot. They enjoy licking the anti-freeze from cars, because they like the salt! Here are a few photographs of these goats in their natural and human-impacted habitats:

Tundra: The Frozen Prairie

This biome circles the world in the highest northern latitude and in the southern hemisphere is found only in the Antartic Peninsula and Islands close by. Here temperatures often reach about -50°F in the winter. Tundra covers about one-fifth of the Earth's land surface. Because of the cold climate it is impossible for trees to grow, thus leaving room for low-growing plant life and wildflowers. For this reason the Tundra biome looks like a frozen-over prairie land.

In the Ice Age, massive glaciers dwelled here. As the Earth warmed these glaciers retreated leaving bare rock and scoured soils. The freezing temperatures leave deeper layers of soil frozen throughout most of the year- this condition is called permafrost. Only the top layer on the surface is able to thaw out in summer conditions. However, this occurs just briefly because summers are extremely short here. The combination of a harsh climate, the lack of nutrients in the soil, and soil being so sparse make it very hard for any type of plant life to grow.

Despite these harsh living conditions animals still manage to survive here. During the Tundra's brief summer, insects hatch out of eggs which were frozen in the top soil. Creating a vast feeding ground for birds, thousands migrate here during this time to feed on these insects. Millions of migrating waterfowl and shore birds come to the shore and lake areas in the artic tundra of Alaska during the summer months. Also, ravens, hawks, ptarmigans, and the open country owl are common. Besides birds and insects mammals dwell in this icy zone. The Caribou, artic hare, mink, weasel, lemming, wolf, wolverine, brown bear, vole and reindeer, roam this land. And the Polar bear, walrus and artic fox are commonly seen on the ice pack and coastal areas.

Temperate Forests: Seasonal Changes

The temperate forest biome is found in the middle latitudes around the globe and this biome is very seasonal.

Temperate forests dominate the mid-latitudes in eastern North America, western Europe, and eastern Asia. In the southern hemisphere, smaller areas of temperate forest can be found in South America, southern Africa, Australia, and New Zealand. Because temperate forests are highly seasonal they have warm summers and cold winters. The trees being deciduous (meaning they drop their leaves in the fall) change colours as the seasons cycle: the green leaves of summer give way to the grey bare branches of winter. Temperate forests blend into the pines and firs of the taiga (cold climate forests).

The Temperate forest biome is one of the most altered biomes on the planet. By looking at a map you will see that our population density very closely corresponds to the distribution of Temperate Forests. We use the wood of these trees for construction, firewood and art. They have been cleared for farming and to build communities. These human activities have led to the decline and loss of these forests in many parts of the world.

MARINE/ISLAND

Islands vary tremendously when it comes to shapes, sizes and climates. Some are tropical and lush, while others are jagged and arid. They also support thousands of unique and unparalleled types of animals, such as the giant tortoises of Galapagos, Madagascar's lemurs, and the humongous Komodo dragons, the largest lizards in the world. Oceanic and continental are two broad categories of islands, regardless of the islands' shape, size or the type of animals that dwell on their surfaces.

Oceanic islands were created from the lava of giant underwater volcanoes. An example of this would be the Hawaiian Islands. Normally these islands are located far from major land masses. Continental islands once were part of a larger land mass and in recent geologic history, have become separated from the mainland because of rising sea levels after massive glaciers melted thousands of years ago or by earthquakes which seperated them from the original land mass. An example of a continental island would be California's Channel Islands. These types of islands have an older and more complicated geological history than the oceanic islands.

The Desert: Land of Little Rain

Of all the biomes of the world, the desert biome has the driest climate.

The great expanses of the world's desert lie between 20 degrees to 30 degrees north and south latitude. It is here that equatorial air falls down toward the Earth's surface and rainfall is rare because rain usually occurs when air begins to rise, not fall. The equatorial air that is falling prevents most air from rising. North Africa, southwestern North America, the Middle East, and Australia support the largest deserts, but there are smaller deserts in other regions such as on the Pacific coast of South America (the Atacama) and the Atlantic coast of southern Africa (the Namib), where moisture from cold water currents is evapourated immediately by the hot land masses adjacent to the currents.

Rainfall in the desert often totals a few inches yearly or, in some regions, there is absolutley none. Desert soils are often salty because whatever little rain that does fall quickly evapourates from the ground, leaving salt and other minerals behind.

Since rainfall is so scarce, plants in the desert are almost always drought-tolerant, meaning they can survive without water for a long time. With unique features such as, thick or waxy leaves, large root systems, and water storage systems-like in the cactus, these adaptive plants are built to store water, find water quickly or live with the littlest amount of water possible. The vegetation in deserts varies tremendously. The mojave desert in California is known for it's unique Joshua trees, in the Sonoran Desert there is the thorn-covered Ocotillo and the giant Saguaro Cactus. Sagebrush covers the Great Basin in the western United States and the Chihuahua Desert (in between Mexico and Texas), is well known for its mesquite trees.

The animals which live in this arid biome are usually light-coloured and use camouflage to blend into their surroundings, and possibly for protection against predators. Being more active at night and around dawn and dusk, allows them to escape the scorching heat. During the day they often lay in burrows or under rocks. Different species of life in the desert include jackrabbits (North America), kangaroo rats, owls, snakes, lizards and tortoises.

Because of our carelessness, deserts are spreading over regions where there was once green, fertile land. This is mainly due to misuse of the planet. However, the desert expansion can be stopped through anti-desertification projects, and better land and agricultural management. This will turn the deserts back, by making the soil stronger with the roots of plants embedded into them, and replenishing the soil with nutrients and minerals.

Tropical Dry Forests

With a well defined dry season, tropical dry forests have high temperatures throughout the entire year.

A home to such animals like the Tasmanian devil and the American alligator, this biome occurs in areas of slightly lower rainfall next to tropical rain forests. Found primarily in Central America, southern Asia (forests here are known as Monsoon Forests), and some portions of Australia, their well defined dry season limits plant growth and the activity of animals.

There is really no actual distinction between this zone and the tropical rain forest because the length of the dry season varies tremendously throughout the tropics, one biome usually gradually changes into the other over hundreds of miles. Wetter or drier soils sometimes produce pockets of tropical dry forest within a tropical rain forest.

Cold Climate Forests: The Taiga

This cold climate that supports coniferous trees (which means that they carry cones) is found at very high latitudes extending across Eurasia and North America. Rainfall in this climate is moderately high but is spread throughout the course of the year, with snow covering the ground in winter. Very little water is evapourated by the sun, thus ponds, lakes and bogs also known as "muskegs" are found everywhere, especially in glacially carved areas.

Vegetation Found in the Taiga

Trees in the taiga (Taiga is a Russian word) use a lot of energy to grow their leaves, thus they have found a way to keep their needles all year round. This way, when the sun comes out again in the spring these trees are already gathering much needed sunlight instead of wasting more energy to grow new leaves. In addition they have adapted their needles to be filled with a chemical that repels grazing animals, and their thick bark resists the loss of moisture in the cold winters. Trees of this biome are also known as boreal or the Northern coniferous forests, usually have shrubs underneath them with blueberries (which is a favourite food of many animals) which act as heath plants.

The days in the Taiga are very short in the winter, as short as six hours. In the summer the days lengthen and plants grow rapidly in the 70°F weather.

Along the river banks throughout the taiga, willows and many other well known trees can be found. Leaves cover the ground for the relatively low temperature and the acidic soil slows down the process of decay.

Taiga Animals

Many animals migrate to the taiga in the summer months. However, those who do not have learned to adapt to the cold.

Moose, wolves, woodland caribou, wood bison, black bear, marten, lynx, and the arctic ground squirrel are common, although they are not as abundant as the mammals living in the grasslands and the savanna biomes. Most of the animal activity in the taiga is seasonal, with large quantities of birds, such as the redpoll, raven, gray jay, red-throated loon, northern shrike, sharp-tailed grouse, and fox sparrow, present only in summer. Also the bald eagle, peregrine falcon, and osprey which are fish eaters, live in this biome.

For the animals that stay in the taiga during the winter months, conserving heat is one of the most important steps of survival. Most animals go into long-term hibernation and other animals such as the Canadian Lynx grow an insulating layer of fur or in other cases feathers. In order to conserve heat some animals have a rounded body structure, with shortened limbs to create less heat loss from long limbs and skin surfaces. Also other animals grow fur or plumage that camouflage with the snowy white background.

THE GRASSLANDS

Throughout the world, grasslands are known by many different names. In North America they are known as "prairie", in Asia "steppe", in South America the "pampas" and in South Africa a "veldt".

This biome is a highly seasonal environment of temperate regions. Grasslands stretch basically thousands of miles primarily in the continents of North America and Asia. A limited area of grasslands lies across southern South America. Its hot summers cause the flora to be baked into a nice crisp golden brown, and in the winter the land freezes over and is carved by powerful winds that are able to gather terrific speed because of the vast open area grasslands provide.

The seasonal changes always have helped maintain this biome. However not just the seasonal changes can keep this biome in check. Wildfires help maintain this biome.

Rainfall in the grasslands is somewhere between the amount of precipitation temperate forests and deserts receive.

However, the amount of rain is not sufficient enough to support clusters of trees. Regions where more rainfall falls than another area results in taller grasses that can reach impressive heights. These grasses dominate the vegetation in this biome. Besides the grasses that grow here, wildflowers every spring provide a beautiful display of colour. Arriving when this biome's rainy season comes, wildflowers help transform the land from a dreary solid brown into a green rainbow-filled landscape.

Wildfires play a very important role. They allow the grasslands to be open and free of trees and shrubs. Any seedling of a tree that appears is killed off by the intense heat. If the fire does not clear out the seedlings, the grasslands would become a shrubland changing forever. Remarkably, the plants in the grasslands have adapted to the wildfires that come through, and actually need them to keep healthy and grow new vegetation in the spring. Not only do wildfires reduce the growth in this area but so do grazers such as, bison, deer, and horses.

Savannah: Tropical Grasslands

Dominating the continent of Africa, savannahs are also found in India and the northern part of South America.

Savannahs are in fact tropical grasslands for they are located at tropical latitudes, however much drier than many tropical forests. Rainfall in this biome is between 20 to 60 inches a year, and can be very seasonal (usually falling within a time period of a couple weeks). Growth after the rainfall occurs, however long periods of drought follow.

Throughout the savannahs, the dominant plant life are grasses and small plants. Trees are sparse throughout this semi arid landscape, only growing where there are cracks in the surface or deep soil. In many savannahs around the world palm trees play an important role in the landscape. The most dominate wooded form in the savanna are the thorn woodlands. Often following the thorn woodlands come tropical dry forests. There is a large amount of wild fruit-trees, which provide food for many birds and animals.

Savannah Animals

Types of birds found in the Savannah biome are shrikes, hornbills, grey louries, flycatchers, knysna, purple-crested louries, green pigeons, rollers and raptors. Larger mammals of this biome are lions, leopards, cheetahs, elephants, buffalos, rhinoceroses, giraffes, hippopotami, gazelles, zebras, kudus, waterbucks, oryxes and many others.

Tropical Rainforests: Equatorial Forests of Rain

Of all the world's forests, it is those in the tropics that face the greatest threat from mankind. Tropical rainforests are one of nature's treasures, and many of them are now at risk. We have already destroyed half of the world's original tropical rainforests! Just in a few decades, we can possibly witness the complete elimation of the world's rainforests. The biodiversity of this biome is legendary — this biome contains the largest biomass. Did you know that enough rainforests are being destroyed every minute to fill 50 football fields? We need to preserve these valuable resources because they are the lungs of our planet, and can possibly hold cures for many of our most deadly diseases. The tropical rainforests are a critical link in the ecological chains of our our earth's biosphere

- Amazon rainforests produce about 40% of the world's oxygen
- One in four pharmaceuticals comes from a plant in the tropical rainforests
- 1400 rainforest plants are believed to offer cures for cancer

40% of tropical rainforests have already been lost in Latin America and Southeast Asia,, Technically, this type of forest can be defined as a forest in the tropics receiving 4-8 metres of rain each year. Tropical rainforests are found in Central and South America, Southeast Asia and islands near it, and West Africa. There are smaller rainforests in northern Australia and other small islands. All tropical rainforests are found along the equator where the temperatures and the humidity is always high, with the days being equal to the nights.

Generally, there is a flow of air that comes from the poles of the Earth, towards the equator. These winds are filled with moisture and the intense heat that is located at the equator causes the moisture to rise, cool and then condense to create rain. This continuous cycle causes it to rain almost 24 hours a day around most of the tropical areas. In some regions there can be more than 15 feet of rain a year. There are one or more "dry" months in this tropical biome, however if one would visit they would see it still is astonishingly wet.

Important Facts: Despite covering only 2% of our planet's surface, over half of the earth's animal, insect species, and flora live there.

Within a four mile square area of a tropical rainforest, you would find:

- Over 750 species of trees
- 1500 different kinds of flowering plants
- 125 species of mammals
- 400 species of birds
- 100 reptiles
- 60 amphibians
- countless insects
- 150 species of butterflies

What is an Ecosystem?

The term ecosystem was coined by the British ecologist A. G. TANSLEY in 1935. He used it to define a unit that covers all organisms of a given area as well as their relationship to the inorganic environment. The organisms within an ecosystem form a biocenosis, their inanimate environment is called a habitat. The totality of all ecosystems on earth is called the biosphere.

Ecosystem is an operational term, since it describes units that are less easily captured than, for example, a molecule, a cell or a species. What is defined as an ecosystem is within the discretion of the person dealing with the ecosystem. A lake, for example, a reed belt, a wood, or a cornfield can be defined and described as ecosystems. Each system can consist of a number of partial systems as can be deduced from the system theory. Which level of complexity is studied is therefore mainly dependent on practical considerations.

It is distinguished between simple and complex, between aquatic and terrestrial, between natural and human-influenced ecosystems. Among the astonishingly complex systems, i.e. those systems that are especially rich in species are tropical rain forests and coral reefs.

Systems are more than just the sum of the system elements' performances. Between their elements exist numerous, often specific and usually always regulated connections. Controlled systems, in contrast, are characterized by feedbacks and thus by a high degree of stability. They are largely independent against disruptions. The higher the number of system elements and the amount of interactions with each other, the more effective fluctuations can be balanced out. But still each system has only a limited capacity. When the limit of this capacity is exceeded, then the system will not return to its original situation or will even be destroyed. Ecosystems with only a few system elements are extremely liable to break down. Just think of a spruce monoculture or a corn field. They can only be kept in balance with the use of stabilizing, energy-consuming measures like the use of insecticides. They can, on the other hand, easily be rebuilt after complete destruction. Complex systems are very resilient, they have a large buffer capacity, but their destruction is an irreversible act. A destroyed tropical rain forest or a destroyed coral reef is lost forever.

Natural or almost natural ecosystems were able to absorb human influences for centuries, but it becomes clear that the limits of the strains on them are reached or even exceeded. Ecology has therefore during the last decades been in the public eye. Environmental protection has become a major issue of the domestic policy of many industrial nations. Many students of biology study this subject only, because they are interested in environmental issues.

Ecology cannot solve political problems, it is especially unable to offer a fast aid to decision-making in order to answer short-term questions. Changes in a human society require a knowledge about its structure, its functions, its decision-makers, the chances to alter something, and the most effective ways to initiate changes. Successes that can be achieved by organizations like Greenpeace helped considerably to create an environmental awareness among general public and politicians.

Ecology is a holistic science. In ecological work, long-term experience, a detailed insight into plants, animals or microorganisms, a sound knowledge and/or mastery of physical and chemical measuring are more important than spontaneous political activity. A co-operation between scientists of different disciplines is therefore the best way to approach the aim. A solid scientific management is necessary in order to co-ordinate the activities of all scientists participating in the project.

Ecosystems change permanently. Changes and disruptions are not only caused by human activities. The evolution of organisms and their adaptation to variable environmental conditions are the main reason of the changes and thus also of the evolution of ecosystems. A sequence of differently structured systems under constant abiotic conditions is called succession. In Central Europe, for example, certain types of forests are the most stable systems. Such a system is called a climax association. Such final links of developments exist only as long as the environmental conditions are stable. In the history of the earth, several drastic climatic changes occurred that caused the extinction of whole groups of plants and thus also to the replacement of once stable ecosystems by others. The pre-Ice Age vegetation of Central and Northern Europe, for example, or the carboniferous swamp forests became irreversibly extinct. After the ice had gone, the landscape was inhabited again beginning in the south and slowly advancing towards the northern part of Europe. A comparison of today's Central and North European flora with that of North America shows that the European flora is far poorer in species. The variety of species of North America that was only to a small extend covered by ice the variety of species that existed before the Ice Age in Europe, too, remained.

The transformation of the natural landscape into a cultivated landscape had a serious impact on all ecosystems. Often, new ecosystems possibly worth protecting developed that could not exist under natural conditions. The Calluna vularis areas of the Lüneburg Heath in Germany are impressive examples. Without regular grazing by a species of German moorland sheep called Heidschnucken, these areas would soon be replaced by birch-pine-forests.

How to Analyze Ecosystems

Ecosystems are usually stochastic systems, on one hand, because they depend on a variety of different parameters, some of which, but never all can be captured. On the other hand are they stochastic, because they are always open systems. Their most important members, the organisms, depend on a steady supply of energy. Two conceptionally different methods to gain information about ecosystems and biocenoses exist.

Stocktaking:

Capture of (all) species of a habitat,

Determination of the percentages of the single species,

Determining of changes in the species composition as a function of time (a. during the cycle of the year, b. over a longer period of time),

Determination and capture of all those chemical and physical parameters, which the organisms depend upon.

Stocktaking has a long tradition. Mapping of vegetation and plant geography are just two methods to be mentioned here. You can find more about stocktaking in the topic plant societies.

Systems Analysis

The attempt by systems analysis is younger than stocktaking. The advent of cybernetics during the 1940th resulted in aids that help defining systems and system properties mathematically in order to develop models suitable for predictions.

Both attempts - stocktaking and systems analysis - complement each other. Only if the results gained by both methods are taken into consideration, it can be tried to understand an ecosystems and to recognize its balance, its behaviour, and its sensitivity towards disruptions as well as to comprehend future developments.

The fundamental principles of the system theory are discussed in the chapter about cybernetics.

1. The system elements, in this case the producers, the consumers of the first order, the consumers of the second order, and the decomposers are arranged one after the other. The arrangement is hierarchical. The single levels of hierarchy are called trophic levels.
2. The system elements are interconnected by a flow of energy and material. The flow of energy is a linear process, the flow of material is a circuit.

Only plants and a few microorganisms are able to convert light energy into chemical energy. They are therefore called producers or primary producers, P. All other organisms are consumers of the first, second or third order or dor decomposers, also called saprophytes. The life-style of producers is called autotropic, the life-style of consumers is heterotropic. A mathematical description of an ecosystem requires four basic elements:

1. System variables (v_i) are a group of variables (v_1, v_2 ... v_n) that describe the state of the system at a given time. Among them are, for example, the biomass (in gram, kilogram, or tons dry weight per square or volume unit). v_1 describes thus the producers (P), v_2

the consumers of the first order, v_3 the consumers of the second order (carnivores, for example).

2. Transfer functions (F_i) are equations that describe the material turnover of a system. They cover, for example, that part of the biomass that is lost by respiration or that part that is used to feed the organisms at the next higher trophic level. The changes of a system variable as a function of time can be described by this differential equation:
   $dv_i/dt = f(v_1, v_2 \ldots\ldots\ldots vn, F_1, F_2 \ldots F_n)$
3. Inputs are amounts of energy or material that are available for a system, like the amount of useable sun energy, the limiting amount of minerals in the soil or the temperature as a factor determining reactions. If only the transfer functions from one trophic level to the next higher one are of interest, then the corresponding F_i becomes the input. The amount of herbivores, for example, is decisive for the amount of carnivores.
4. Proportional factors (c_i) and constant factors are unchangeable. Among them is the amount of food an animal needs per period of time.

The mathematical model of a system can now be given by a set of equations that describe the flow of energy and material between the single levels. Matrix calculation is used in order to sum up sets of data and equations. A matrix is a two-dimensional list of data, where each value can be described by its co-ordinates i and j. Matrices can be multiplied with each other. The result is a new matrix. Such a mathematical description shows, for example, which system component has a direct influence on another one. In our generalized ecosystem, this would result in an image that shows that each system component has an effect on itself, but that carnivores, for example, have no direct influence on plants.

Ecological research produces always large amounts of data. It does therefore make sense to make them suitable for the computer, i.e. to group them right from the beginning into matrices that can easily be processed by the computer. No exact sets of data can be collected for many of the required parameters. The inputs fluctuate usually very much, due to unpredictable weather conditions, for example. The number of interactions occurring in a natural ecosystem is usually bigger than those given in its model. Animals cannot be clearly grouped into herbivores or carnivores as many of them are omnivores. But if such complications are recognized, it is relatively easy to integrate them into an existing system. Often, nevertheless, the necessary information is lacking. On the other hand, relations exist that have hardly any quantitative input. It does make sense to not consider them in the model in order to decrease the amount of work.

Until now, very few ecosystems have been described that show the properties of the system exemplarily. In Germany, there is a lake, the Plußsee in the vicinity of Plön/Holstein, Germany, that can be cited as an example of an almost completely explored ecosystem (the work was done by the Max-Planck-Institute of Limnology at Plön, Germany). The "Solling-project" is an example of a terrestrial ecosystem, a wood, that is analyzed by researchers at the University of Göttingen, Germany.

The majority of ecologists have a solid knowledge of plant and animal species, but know only little about soil bacteria and fungi (decomposers). Hardly any data about their species composition, their population density, and their rate of multiplication under natural conditions do therefore exist. Kinetics of multiplication that were taken under lab conditions are only of partial use as those growth conditions are usually better and the rate of multiplication is thus higher than under natural conditions.

Attempts of system analysis can optimized so much that little and incomplete information becomes sufficient to make relatively apt predictions. Ecology depends on optimization processes comparable to those of projections done during elections, since all lists of data however extensive they may look represent always just a relatively small spot check of the actual not yet captured data. The accuracy of a mathematical model increases with increased complexity, while an ecosystem becomes more stable the more complex its is. Statistics shows that a systematic mistake (s) is inverse proportional to the root of the number of single measurements (1/roots of n). The larger the amount of data, the smaller is consequently the error rate.

THE FLOW OF ENERGY IN ECOSYSTEMS - PRODUCTIVITY, FOOD CHAIN AND ATROPHIC LEVEL

As was explained tha the flow of energy is an essential feature of every ecosystem, since all living systems are open systems. They depend on a steady supply of energy in order to keep up the structural organization and all life-preserving functions. According to the second law of thermodynamics, each system strives for the state of highest entropy. The inversion, i.e. the coming into being of systems that are poor in entropy or are, in other words, rather well organized, requires a steady supply of energy from the surrounding. Since only plants are able to use light energy, they have a key position in every natural ecosystem.

The amount of biomass decreases drastically from one trophic level to the next one as the flow of energy is a directed process and as an optimal ten percent of the biomass of the previous tropic level can be used in the next higher one. As a consequence, four, at the most five trophic levels exist in nature. The biomass of the sum of all carnivores is always smaller than that of the herbivores, and that of the herbivores again is smaller than that of the plants.

The result is a food pyramid, whose shape depends on the productivity and the species composition of the respective ecosystem. The linear succession of the single elements (producers, consumers of the first order, ...) is called the food chain. The conditions in natural ecosystems are nevertheless mostly more complex than that, so that food web is a better way to describe the actual state. Food pyramids are usually based on biomass, not that often also on the number of species or individual organisms.

More than 90 percent of the total biomass of the earth is made up by plants, only a few percent are allotted to all other groups of organisms. A comparison of the number of existing species in contrast shows that about ten times more animal than plant species exist. Under certain conditions, inverse food pyramids occur. The reason is either a spatial or a time shift in the appearance of the single system elements.

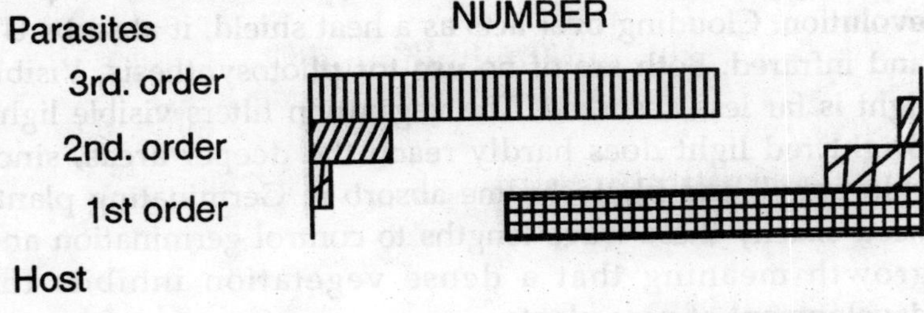

Fig. Inverse Pyramids

The number of parasites is larger than the number of hosts. This is due to the fact that the parasites are far smaller than the hosts, so that every host harbours several parasites. A number of animals live from dead plants or plant remnants. The biomass of living animals is therefore in some winters higher than that of the living plants. The relation is just the other way round during all other seasons.

A spatial separation is typical for the deep sea, because deep sea animals live in depths were no photosynthesis and thus no plants can occur anymore.

These animals live from a constant 'rain' of dead plants, almost all of them single-celled algae. In order to determine the size of the energy flow in an ecosystem, the input of energy, i.e. the amount and quality of the sunlight have to be determined first.

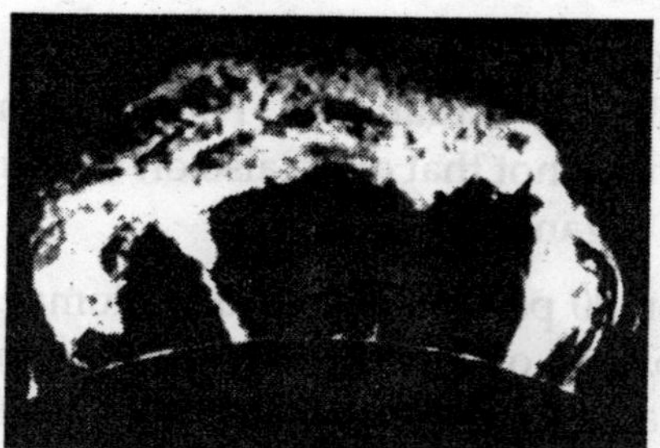

Fig. The Emission Spectrum of the Sun and the Filter Effect of the Biosphere

Those pigments that make optimal use of the available wave length of the light were selected in the course of plant evolution. Clouding over acts as a heat shield, it absorbs UV and infrared. Both are of no use for photosynthesis. Visible light is far less absorbed. The vegetation filters visible light. Bright red light does hardly reach the deeper areas, since chlorophyll and phytochrome absorb it. Germinating plants need exactly these wave lengths to control germination and growth meaning that a dense vegetation inhibits the development of new plants.

The average energy of the sun of the whole earth's surface is 2 calories (cal)/square centimeter (cm^2)/minute (min). It is also called the solar constant. Seasonal fluctuations as well as differences in the exposition (southern or northern slope) cause differences in the order of magnitude. The solar energy reaches thus a mean value of 3,000 - 4,000 kilo calories (kcal)/m^2 (120,672 kilo joule (KJ)/square metre (m^2)/day) or 1.39 KJ/m^2/second. A part of this energy is reflected by the earth surface and is thus not available for biosynthetic processes. The remains are called the net radiation that is 1 million kcal/m^2/year over the sea between 40° north and 40° south. The net radiation over land is 0.6 million kcal/m^2/year.

Evapouration and movements of the air are important factors that cause the largest part of this originally enormous amount of energy to be given off into space. Without this, the earth would rapidly overheat and life would become impossible. On the other hand, the sun energy and the resulting increase in temperature are the main reasons for the existence of climatic zones as well as of seasonal and daily fluctuations.

The yearly production of biomass is an estimated 164 billion tons. A third of it is produced in the oceans, two thirds in terrestrial ecosystems.

Biomass and rate of production are two different things. The rate of production or productivity means an amount of energy bound in a certain period of time. Existing biomass alone allows only under certain conditions to determine approximate values of the turnover of energy. Each step in metabolism loses energy as heat. Therefore, it is distinguished between gross and net production. The net production is the amount that remains after the energy that is lost due to respiration has been subtracted.

Among the difficulties of estimating the productivity is the fact that it can often not be distinguished whether a system (in this case a plant) is in a steady state or in a growth phase. A continuous decrease of growth results in a decrease of respiration, too, because the maintenance of structures requires lesser energy than their production.

The determination of the photosynthetic activity that is measured as the amount of emitted oxygen tells little about the productivity as a part of the oxygen is consumed by respiration and another part is required for photorespiration. The activity of photorespiration is directly correlated to the amount of available light. The single factors can be captured separately under controlled lab conditions and with a large experimental effort. From the data captured under these conditions, proportional factors can be determined. The biomass is often ascertained by measuring the amount of carbon. The factors of conversion are:

10 kcal ~ 2g dry substance ~ 1g carbon

The flow of energy is usually given as kcal/gram (g) dry weight. Natural ecosystems are optimized for a high turnover, while ecosystems influenced by humans, especially agricultural areas, are optimized for an as high as possible rate of net production.

Climax associations, like the tropical rain forest are characterized by a high productivity at an almost constant biomass. An ecosystem that is in a succession is in a very different state. A bog, for example, grows by the continuous deposition dead Sphagnum.

Only about 5 percent of all available sun energy is conserved as chemical energy in the biomass of plants. A theoretically optimal 80 percent of this energy can be used by the organisms of the next higher trophic level. In reality, the amount is usually far smaller. The major part of the biomass of woods, for example, is made up of wood that is extremely unfavourable for animal feeding. Only a few specialists can use it. The relation between gross and net production as well as the flow of energy through the system can be demonstrated especially well with the model system wood.

The available amount of light is usually no limiting factor. This is different with the supply of water. Just think of the meagre plant growth in deserts as compared to the lush vegetation of the tropics. This difference is certainly not due to light as the amount of light in the usually cloudless deserts is far larger than that of the clouded tropics. Irrigated desert areas yield a higher gross production than areas with a lower light intensity. The high losses due to respiration during warm nights lead to a higher energy consumption than in cooler areas. The rate of net production is thus lowered. This explains, why the yield per hectare of rice harvests in equatorial areas is always lower than that in temperate zones.

The net primary production may fluctuate considerably. The lowest value for 'open sea', for example, is two kcal/g, the highest is 400, the lowest net primary production of cultivated areas is 100, the highest is 4000. The values fluctuate roughly fivefold in most other types of vegetation. In order to calculate the energy fixation from the net primary production, a calorific value (kcal/g) is used. It is 4.5 (kcal/g). Grassland has the lowest value (4.0 kcal/g), while the open sea has 4.9 kcal/g.

NUTRIENT CYCLES

The elements required for living systems are mostly stable. Radioisotopes play a minor role, and even the half-value period of 5770 years of ^{14}C is much higher than the life of the average organism. As a consequence, all elements of a biosphere are preserved. All that changes is their distribution within the biosphere, some are taken to new places, others are chemically integrated into different molecular components.

All changes can be described by cycles. Cycles can be drawn up for any element, but in biology, that of oxygen, carbon, nitrogen, phosphor, sulphur, as well as water are of interest. Water is no element, but a largely stable molecule that is essential for all processes of life. Cycles can be described as systems just like the generalized ecosystem that we just got to know. Often, large pools of elements like the earth's crust, the ocean or the atmosphere are the system elements. Only a minute part of the material mills around and even less participates in the development of living systems.

THE WATER CYCLE

The water cycle is the first to be introduced as the water supply of the earth's surface is besides the sun and the amount of energy provided by it the most important precondition of vegetation and thus also of the colonization of the earth. This known fact becomes especially impressive through satellite takings made of partially irrigated dry areas.

Water can exist in three different states: solid (ice), fluid, and gaseous (vapour). It is in its fluid state at temperatures that are optimal for physiological processes. Water has a high heat storage capacity. It is moreover an ideal solvent for numerous ions. The pH-value of underground water fluctuates between pH 3 and pH 10. It is high in the coastal areas of the oceans and still higher in the open ocean. The high alkalinity is one of the main courses of the water's high capacity for carbon dioxide that becomes carbonate/bicarbonate in the water. Only a fraction of the existing water is altered chemically. The photolysis taking part during photosynthesis makes up the main part of this fraction.

The earth surface is 510 million square kilometers, 362 million of it are covered by water and 325 million square kilometers of it are open oceans. The total amount of water is 1.5 billion square kilometers. 97 percent of it is ocean water. Only three percent is freshwater, and three quarters of it is immobile ice that makes up polar icecaps and glaciers. Its atmospheric part is lower than 0.001 percent. The amount that is available for plants has roughly the same magnitude.

Atmospheric water has a fluctuating geographic distribution. It occurs mainly in the equator region. Water remains an average 9-10 days in the atmosphere, though it can return to earth after a few hours or after weeks. The atmospheric differences in temperature and pressure are the main cause of air circulation. Above the oceans, less water comes down than enters the atmosphere (107-114 centimeter (cm)/year as compared to 116 - 124 cm/year). The situation is just the opposite above land. Here, 47 cm/year evapourate into the atmosphere, while 71 cm/year precipitate. The difference is balanced out by the draining away of surface water, especially rivers, and to a lesser extend by underground water. Two further aspects are important for a botanist: the difference in the geographic distribution of the rainfall and the part of the cycle that the plant itself takes part in.

The production of 20 tons of biomass requires 2,000 tons of water. The largest part of the water does thus pass the plant leaving it by way of transpiration. 15 of the 20 tons of fresh weight are made up by unbound water within the plant tissue. The other five are dry weight. Three of them are bound water, the rest consist of other substances.

THE OXYGEN CYCLE

The oxygen of the atmosphere is almost completely the product of the photosynthetic activity of green plants. The oxygen content of the atmosphere increased continuously during earth history since the first occurrence of organisms that were able to split water and perform photosynthesis. The increase stopped when a balance, today's 21 percentage by volume was reached.

The source of the atmospheric oxygen was mainly water and to a lesser extend other oxides. Respiration and combustion are energy-consuming processes. The end product of respiration is carbon dioxide. The atmosphere contains about 1.3 x 10^{14} tons of free oxygen. The lithosphere carries even 5.5 x 10^{16} tons of bound oxygen, more than a hundred times more than the atmosphere. Its main part is bound as carbonates, silicates, sulfates, and other oxides. In the atmosphere, oxygen occurs mainly unbound, while the strongly ionizing cosmic radiation causes the production of ozone and atomic oxygen within the stratosphere. The ozone layer protects the biosphere effectively from short-wave UV-radiation.

Over the decades, human activities have used up increasing amounts of oxygen while carbon dioxide is set free. A decrease of free oxygen is still very unlikely. A noticeable increase in the atmosphere's concentration of carbon dioxide can nevertheless be registered.

A main characteristic of the oxygen cycle is its connection with a part of the carbon cycle that plants have a key position in. The exchange rate of atmospheric oxygen is a relatively high: 2,000 years. In other words: plants produce 1/2000 of the whole atmospheric oxygen each year and the same amount of oxygen is consumed by oxidation. The carbon dioxide of the atmosphere is completely exchanged within 300 years. The total amount of water, namely the already mentioned 1.5 billion cubic kilometer is completely split by photolysis and reproduced by oxidation within 2 million years. These values show the water cycle changes just little without the activity of plants, while the oxygen and the carbonate cycle would be altered drastically without plants.

THE CARBON CYCLE

Carbon is an element that occurs for the smallest part as atoms at the earth surface (carbon, diamonds). Its main part is either oxidized (carbon dioxide, carbonate/bicarbonate and a small amount of carbon monoxide) or reduced (hydrocarbons and their derivatives). The main part of the carbon of organic substances is reduced, and the fixation of carbon occurs during photosynthesis. The plants have thus a key role again. The carbon cycle is directly linked to their energy budget.

Often, a subsection of the carbon cycle, the carbonate cycle, is excluded and discussed separately. This may be practical as the turnover of carbon dioxide and water-dissolved carbonate/bicarbonate is especially easy to determine. The ^{14}C-method helped essentially to understand the whereabouts of the carbon.

In the carbonate cycle, the reduced state of carbon is usually treated as a 'black box'. We do indeed know least about it. It contains the living substance (biomass) but also dead material that is often collectively called fossil fuel. The amount of the carbon fixed by plants is directly proportional to the gross primary production, while the amount of carbon dioxide that is set free by respiration depends on the difference between gross and net production.

The main part of carbon dioxide exists as carbonate in the lithosphere. A smaller part is dissolved in the oceans. Carbon dioxide comes to 0.03 volume percent, i.e. 300 parts per million (ppm) of the atmosphere. This is a guideline. The carbon dioxide concentration increases continuously since the late 1950th. The function given in the following illustration can be extrapolated backwards showing that the carbon dioxide concentration must have been 260 ppm in pre-industrial times. The carbon dioxide content of the air increased about 8 percent between 1958 and 1982. This increase was caused by the burning of fossil fuels like coal, oil, and natural gas.

Based on details from 1979 and 1980, the amount of carbon dioxide can be calculated up to 5.3 billion tons per year. A further 1,8 - 4.7 x 10^9 tons of carbon set free as the result of the destruction of the biosphere have to be added. Among this destruction is the clearing of the tropics, the destruction of the savannas as well as frequent plowing of cultivated areas. Breaking up the soil causes carbon dioxide that was produced as the result of the decomposition of organic material to surface, its absorption to humus particles does not take place, the carbon dioxide escapes into the atmosphere or is washed away into rivers and finally into the sea. The numbers given for the atmospheric carbon dioxide concentration are average values. The concentrations may fluctuate considerably regionally and locally as a measurements close to a patch of forest show. Besides day-night fluctuations different values were captured at different heights above the soil.

The lowest concentration, 305 parts per million (ppm) is given in turquoise, the highest concentration, 350 ppm, is given in dark blue. The vertical distribution of carbon dioxide in the air around a group of trees fluctuates with the time of the day. Photosynthesis is shut off at night and as a consequence the respiration from the soil can raise the concentration of carbon dioxide at ground level to as much as 400 ppm, while the CO_2 concentration at treetop level can drop to 305 ppm at noon owing to photosynthetic uptake.

Some quantitative data have to be considered in order to understand the global shifts of carbon. Its total amount is 1.384 x 10^{18} tons, 3.9 x 10^{13} of which are inorganic and 1 x 10^{12} tons are organic. The total biomass contains 5.6 x 10^{12} tons of carbon, its annual gross primary production covers 1.1 - 1.2 x 10^{11} tons of carbon, the net primary production is 0.57 x 10^{11} tons of carbon. About half of the gross primary production, 0.43 x 10^{11} tons of carbon exactly, is made up by marine plants, usually single-celled algae. In other words: the main part of the carbon fixed by primary production is respired either directly or during the biological decomposition of dead organic material.

The lifetime of marine organisms lasts weeks, that of terrestrials years. The expanded lifetime and the associated accumulation of fixed carbon results in a decreased respiratory activity. Only 30 to 40 percent are respired.

The given numbers and especially the increase in atmospheric carbon dioxide and the high percentage of the biosphere's destruction lead to the assumption that the carbon cycle is not in balance any more and that this results inevitably in an increased temperature (greenhouse effect). The reality is far more complex since only a small fraction of the carbon dioxide that is set free during combustion enters the atmosphere and remains there. A considerable part is absorbed by the oceans. There it is used for the building up of the organisms' skeletons. The importance of this process is demonstrated by the existence of extensive chalk layers and the fact that an enhanced chalk formation has been the immediate cause for the name of a whole geological era.

Hardly any hints that the annual production of biomass decreases exist despite the destruction of the vegetation by man in our time. The contrary seems to be the case, maybe because carbon dioxide has always been a limiting factor of photosynthesis. This facts should nevertheless not be overly interpreted as too high concentrations of carbon dioxide cause severe growth disorders as lab experiments showed.

The carbon turnover of terrestrial organisms can only be understood if lithosphere, surface water, atmosphere, and oceans are included. The oceans nevertheless proved to be almost closed partial cycles.

The amount of carbon exchange between ocean and atmosphere are very small. Despite the large mass shifting of the surface water more commonly known as currents the water of the deep sea is hardly affected. Measurements of its ^{14}C-content showed that below 1,500 metres depth the water remains in the Atlantic for 275 years in the same place.

THE NITROGEN CYCLE

The nitrogen cycle is far more complex than the cycles you got to know until now. 79 percent of the atmosphere consist of free nitrogen and at least the same amount of bound nitrogen is found in the lithosphere. These large reservoirs are not immediately available for plants. In this context, microorganisms have a central role. Nitrogen fixation is the cue. This process has already been discussed and you may remember that it is extraordinarily energy-consuming. Plants use nitrogen almost only as ammonium and nitrate ions. In organic materials, nitrogen is mainly required for the production of the amino-groups found in proteins or nucleic acids. Nitrate and nitrite bacteria convert the amino groups back to nitrate or nitrite. Nitrate-reducing bacteria living in the soil or in water reduce oxidized nitrogen compounds and do thus close the cycle. Nitrogen fixation and reduction do roughly balance one each other.

The production of ammonium compounds and nitrates is a limiting factor of plant growth. It is true that the lithosphere contains almost unlimited amounts of nitrates, but they occur mainly in depths that are unreachable for plant roots. For humans, too, it is uneconomic to exploit this pool.

Nitrogen compounds are usually very water soluble, large amounts are therefore lost by leaching. They may, especially if occurring in addition with extensive fertilizing, accumulate in lakes or ponds and cause eutrophication. You will find more about this topic subsequent to the section about the phosphorus cycle.

Many nitrogen fixing bacteria and blue-green algae are free-living, others live in symbiosis with plants like with the leguminosae Cycas or Ginko. The symbiotic species bind about ten times as much nitrogen as the free-living. Free-living bacteria and algae fix on average 1g/square metre/year. The highest measured value is 20 g/square metre/year.

The relatively high rice yields of South and South-East Asia are partly based on the occurrence of extensive populations of blue-green algae like Nostoc. They live in the shallow stagnant waters where rice cultures are cultivated. In the whole contemplation of the nitrogen cycle, global or regional changes are of secondary importance. Far more important are the local concentrations and there again mainly the concentrations in the rhizosphere of the plants.

The nitrogen balance of an ecosystem that is basically self-contained with respect to its nitrogen, an unpastured grassland, is given in the following table. Analyses of this type have repeatedly been made. Without the results gained from them, modern agriculture and modern forestry would not be possible.

Atmospheric nitrogen is mainly inaccessible while nitrogen compounds are often very reactive and toxic. An over-fertilization does therefore often lead to degeneration phenomenons and lower yields instead to a better growth. Nitrose gases are extremely poisonous and thus a main factor in acidic rain.

THE PHOSPHORUS AND THE SULPHUR CYCLES

Of these two elements plants can only use certain groups of compounds: phosphates and sulfates. In the earth's biosphere exist no gaseous and also no reducing phosphor compounds. Gaseous sulphur compounds like hydrogen sulfide or sulphur dioxide are rare and if they occur, they cause damage.

It is not quite right to talk about a phosphorus and a sulphur cycle since the flow of material is almost exclusively linear at least when regarded in dimensions of hundreds or thousands of years. Only when taking human influences like fertilization into consideration, a cycle begins to evolve.

The phosphorus reservoirs are phosphorus-containing rock stratums and deposits of inorganic and organic phosphorus compounds. Phosphates are usually not or hardly water-soluble and therefore unavailable for the plant in this state. The mineral enters the ecosystems by way of a stepwise degradation usually with the aid of microorganisms. The distribution of phosphates differs very much geographically: interesting deposits occur only in Morocco. Today, the prevailing view is that enough worldwide sources of phosphorus exist to supply the agriculture also in the future with enough fertilizer. It depends on the general political climate and on the price whether this optimistic prognosis comes true. Industrial nations can easily fund the import of phosphates, developing countries cannot.

Organic phosphorus sources are the guano hills before the coast of Peru that developed by accumulation of bird excrements and that are determinedly quarried since the 19th century.

In contrast to phosphorus, sulphur is both oxidized and reduced by microbial processes. The plant can only use sulfates. Sulfates and phosphates are water-soluble and are thus easily eroded from the soil. A part accumulates in stagnant waters like lakes and ponds and contributes to their eutrophication. The main part of the phosphates is finally washed into the oceans where it is converted to insoluble compounds that accumulate at the bottom of the sea and are lost to the biosphere for the time being.

SPECIES

This question is important. The scientific system of naming "kinds" of plants and animals revolves around the species level. A scientific name, such as Turdus migratorius (American robin), is always written in italics, and contains first the genus and then the species names. A name like Pinicola enucleator californica (California pine grosbeak) also contains a subspecies name (= californica). For many species (ones that vary in size and/or colour over geography), subspecies names are used for distinctive geographic forms.

For a long time, ornithologists almost universally used the biological species concept (BSC). This definition of "species" is based on species being reproductively isolated from each other. Under this definition, distinctive geographical forms of the same "kind" of bird are usually lumped as one species. This is because the geographic forms interbreed (or probably would, if they had the chance) where they intersect on the map. The problem with this definition is slightly different, geographically-isolated, forms rarely present us with "tests" of their willingness to interbreed. According to to adherents of the BSC, if the forms are only slightly different, they would probably interbreed if given the chance. Thus, they should be considered the same species. However, proponents of the BSC also say that because two things rarely interbreed (and produce viable hybrids) doesn't mean they belong to the same species. For example, wolves and coyotes (there's no educated disagreement that these are different species) can mate and have fertile and healthy pups.

The phylogenetic species concept (PSC) says that diagnosable geographic forms of the same basic "kind" of bird should be treated as distinct species. This is because these forms have evolved separately, and have unique evolutionary histories. The PSC is gaining favour because there is no worry about whether slightly-different geeographic forms might interbreed. If they don't, for whatever reason (for example, migration to different breeding areas), they are full species. Obviously, the PSC is less restrictive than the the BSC. There would be many more species of birds under the PSC than under the BSC.

The debate over how species should be defined will continue as long as people are allowed to think freely. Perhaps the argument is rhetorical, because every kind of organism presents a unique situation. It is possiblew that neither definition can be applied consistently in nature.

Species

- A group of organisms that have a unique set of characteristics (like body shape and behaviour) that distinguishes them from other organisms. If they reproduce, individuals within the same species can produce fertile offspring.
- The basic unit of biological classification. Scientists refer to species using both their genus and species name. The house cat, for example, is called Felis catus.

- The lowest principal unit of biological classification formally recognized as a group of organisms distinct from other groups. In sexually producing organisms, "species" is more narrowly characterized as a group of organisms that in natural conditions freely interbreed with members of the same group but not with members of other groups
- The boundaries of this taxonomic level (the most precise in the hierarchical system of binomial nomenclature) are hotly debated among scientists and there is little real consensus about where to draw the lines between species, subspecies, morphs, races, variants, etc. In general, a species is a group of organisms that resemble one another in appearance, general behaviour, ecological niche, chemical makeup and processes, and genetic structure. Organisms that reproduce sexually are classified as members of the same species only if they can actually or potentially interbreed with one another and produce fertile offspring. It should be noted that some (though quite few) taxonomists believe the species level of classification is frequently invalid and these scientists only recognize classifications down to the level of genus (again, these taxonomists represent a very small minority view).

Genus

- Taxonomic group containing one or more related species
- A group in the classification of organisms. Classification level above the species group. It consists of similar species. Similar genera (plural form of genus) are grouped into a family

Tribe

- A taxonomic category between a genus and a subfamily

Family

- a taxonomic group containing one or more genera;
- a division of classification including a number of genera agreeing in one or a set of characters and so closely related that they apparently are descended from one stem.
- A major category in the taxonomic hierarchy, comprising groups of similar genera. Families are thought by some to represent the highest natural grouping. The Latin names of families usually end in the suffix -aceae (plants) or - idae (animals). Groups of similar families are placed in orders. Large families may be split into tribes.
- 5th rank in Taxonomic system. Kingdom, Phylum, Class, Order, Family, Genus, Species. Family names end in -aceae for plants, eg Liliaceae (lilies), and -idae for animals, eg Macropodidae (kangaroos).Biology - Flora & Fauna.

GENE MUTATION AND MENEDEL'S LAW

A mutation is a permanent change in the DNA sequence of a gene. Mutations in a gene's DNA sequence can alter the amino acid sequence of the protein encoded by the gene.

How does this happen? Like words in a sentence, the DNA sequence of each gene determines the amino acid sequence for the protein it encodes. The DNA sequence is interpreted in groups of three nucleotide bases, called codons. Each codon specifies a single amino acid in a protein.

What is a Gene Mutation and how do Mutations Occur?

A gene mutation is a permanent change in the DNA sequence that makes up a gene. Mutations range in size from a single DNA building block (DNA base) to a large segment of a chromosome.

Gene mutations occur in two ways: they can be inherited from a parent or acquired during a person's lifetime. Mutations that are passed from parent to child are called hereditary mutations or germline mutations (because they are present in the egg and sperm cells, which are also called germ cells). This type of mutation is present throughout a person's life in virtually every cell in the body.

Mutations that occur only in an egg or sperm cell, or those that occur just after fertilization, are called new (de novo) mutations. De novo mutations may explain genetic disorders in which an affected child has a mutation in every cell, but has no family history of the disorder.

Acquired (or somatic) mutations occur in the DNA of individual cells at some time during a person's life. These changes can be caused by environmental factors such as ultraviolet radiation from the sun, or can occur if a mistake is made as DNA copies itself during cell division. Acquired mutations in somatic cells (cells other than sperm and egg cells) cannot be passed on to the next generation.

Mutations may also occur in a single cell within an early embryo. As all the cells divide during growth and development, the individual will have some cells with the mutation and some cells without the genetic change. This situation is called mosaicism.

Some genetic changes are very rare; others are common in the population. Genetic changes that occur in more than 1 percent of the population are called polymorphisms. They are common enough to be considered a normal variation in the DNA. Polymorphisms are responsible for many of the normal differences between people such as eye colour, hair colour, and blood type. Although many polymorphisms have no negative effects on a person's health, some of these variations may influence the risk of developing certain disorders.

MENDEL'S FIRST LAW OF GENETICS (LAW OF SEGREGATION)

Genetic analysis predates Gregor Mendel, but Mendel's laws form the theoretical basis of our understanding of the genetics of inheritance.

Mendel made two innovations to the science of genetics:

1. Developed pure lines
2. Counted his results and kept statistical notes

Pure Line: a population that breeds true for a particular trait [this was an important innovation because any non-pure (segregating) generation would and did confuse the results of genetic experiments]

Results from Mendel's Experiments

Terms and Results Found in the Table

Phenotype: literally means "the form that is shown"; it is the outward, physical appearance of a particular trait

Mendel's pea plants exhibited the following phenotypes:

- Round or wrinkled seed phenotype
- Yellow or green seed phenotype
- Red or white flower phenotype
- - tall or dwarf plant phenotype

Seed Colour: Green and yellow seeds.

Seed Shape: Wrinkled and Round seeds.

What is seen in the F_1 generation? We always see only one of the two parental phenotypes in this generation. But the F_1 possesses the information needed to produce both parental phenotypes in the following generation. The F_2 generation always produced a 3:1 ratio where the dominant trait is present three times as often as the recessive trait. Mendel coined two terms to describe the relationship of the two phenotypes based on the F_1 and F_2 phenotypes.

Dominant: the allele that expresses itself at the expense of an alternate allele; the phenotype that is expressed in the F_1 generation from the cross of two pure lines

Recessive: an allele whose expression is suppressed in the presence of a dominant allele; the phenotype that disappears in the F_1 generation from the cross of two pure lines and reappears in the F_2 generation

Mendel's Conclusions

1. The hereditary determinants are of a particulate nature. These determinants are called genes.
2. Each parent has a gene pair in each cell for each trait studied. The F_1 from a cross of two pure lines contains one allele for the dominant phenotype and one for the recessive phenotype. These two alleles comprise the gene pair.
3. One member of the gene pair segregates into a gamete, thus each gamete only carries one member of the gene pair.
4. Gametes unite at random and irrespective of the other gene pairs involved.

Mendelian Genetics Definitions

- *Allele:* one alternative form of a given allelic pair; tall and dwarf are the alleles for the height of a pea plant; more than two alleles can exist for any specific gene, but only two of them will be found within any individual
- *Allelic pair:* the combination of two alleles which comprise the gene pair
- *Homozygote:* an individual which contains only one allele at the allelic pair; for example DD is homozygous dominant and dd is homozygous recessive; pure lines are homozygous for the gene of interest
- *Heterozygote:* an individual which contains one of each member of the gene pair; for example the Dd heterozygote
- *Genotype:* the specific allelic combination for a certain gene or set of genes

Using symbols we can depict the cross of tall and short pea plants in the following manner:

The F_2 generation was created by selfing the F_1 plants. This can be depicted graphically in a Punnett square. From these results Mendel coined several other terms and formulated his first law. First the Punnett Square is shown.

The Punnett Square allows us to determine specific genetic ratios.

Genotypic ratio of F_2: 1 DD : 2 Dd : 1 dd (or 3 D_ : 1 dd)

Phenotypic ratio of F_2: 3 tall : 1 dwarf

Mendel's First Law: the law of segregation; during gamete formation each member of the allelic pair separates from the other member to form the genetic constitution of the gamete

Confirmation of Mendel's First Law Hypothesis

With these observations, Mendel could form a hypothesis about segregation. To test this hypothesis, Mendel selfed the F_2 plants. If his law was correct he could predict what the results would be. And indeed, the results occurred has he expected.

From these results you can now confirm the genotype of the F_2 individuals.

Thus the F_2 is genotypically 1/4 Dd : 1/2 Dd : 1/4 dd

This data was also available from the Punnett Square using the gametes from the F_1 individual. So although the phenotypic ratio is 3:1 the genotypic ratio is 1:2:1

Mendel performed one other cross to confirm the hypothesis of segregation — the backcross. Remember, the first cross is between two pure line parents to produce an F_1 heterozygote.

At this point instead of selfing the F_1, Mendel crossed it to a pure line, homozygote dwarf plant.

Backcross: Dd x dd

Backcross One or (BC_1) Phenotypes: 1 Tall : 1 Dwarf

BC_1 Genotypes: 1 Dd : 1 dd

Backcross: the cross of an F_1 hybrid to one of the homozygous parents; for pea plant height the cross would be

Dd x DD or Dd x dd; most often, though a backcross is a cross to a fully recessive parent

Testcross: the cross of any individual to a homozygous recessive parent; used to determine if the individual is homozygous dominant or heterozygous

So far, all the discussion has concentrated on monohybrid crosses.

Monohybrid cross: a cross between parents that differ at a single gene pair (usually AA x aa)

Monohybrid: the offspring of two parents that are homozygous for alternate alleles of a gene pair

Remember: a monohybrid cross is not the cross of two monohybrids.

Monohybrids are good for describing the relationship between alleles. When an allele is homozygous it will show its phenotype. It is the phenotype of the heterozygote which permits us to determine the relationship of the alleles.

Dominance: the ability of one allele to express its phenotype at the expense of an alternate allele; the major form of interaction between alleles; generally the dominant allele will make a gene product that the recessive can not; therefore the dominant allele will express itself whenever it is present

Mendel's Law of Independent Assortment

To this point we have followed the expression of only one gene. Mendel also performed crosses in which he followed the segregation of two genes. These experiments formed the basis of his discovery of his second law, the law of independent assortment. First, a few terms are presented.

Dihybrid cross: a cross between two parents that differ by two pairs of alleles (AABB x aabb)

Dihybrid: an individual heterozygous for two pairs of alleles (AaBb)

Again a dihybrid cross is not a cross between two dihybrids. Now, let's look at a dihybrid cross that Mendel performed.

Parental Cross: Yellow, Round Seed x Green, Wrinkled Seed

F_1 Generation: All yellow, round

F_2 Generation: 9 Yellow, Round, 3 Yellow, Wrinkled, 3 Green, Round, 1 Green, Wrinkled

At this point, let's diagram the cross using specific gene symbols. The dominance relationship between alleles for each trait was already known to Mendel when he made this cross. The purpose of the dihybrid cross was to determine if any relationship existed between different allelic pairs.

Let's now look at the cross using our gene symbols.

Now set up the Punnett Square for the F_2 cross. The phenotypes and general genotypes from this cross can be represented in the following manner: The results of this experiment led Mendel to formulate his second law.

Mendel's Second Law

The law of independent assortment; during gamete formation the segregation of the alleles of one allelic pair is independent of the segregation of the alleles of another allelic pair

As with the monohybrid crosses, Mendel confirmed the results of his second law by performing a backcross - F_1 dihybrid x recessive parent.

Let's use the example of the yellow, round seeded F_1.

Punnett Square for the Backcross

The phenotypic ratio of the test cross is:

- 1 Yellow, Round Seed
- 1 Yellow, Wrinkled Seed
- 1 Green, Round Seed
- 1 Green, Wrinkled Seed

Chapter 9

Ethology

Biology has become such a vast research enterprise that it is not generally regarded as a single discipline, but a number do assist in understanding the genetic variation of a population; and physiology borrows extensively from cell biology in describing the function of organ systems. Ethology and comparative psychology extend biology to the analysis of animal behaviour and mental characteristics, whilst Evolutionary psychology proposes that the field of psychology, including in regard to humans, is a branch of biology. The study of biology can be divided into different disciplines

ETHOLOGY

The study of animal behaviour. Modern ethology includes many different approaches, but the original emphasis, as expounded by Konrad Lorenz and Niko Tinbergen, was placed on the natural behaviour of animals. This contrasted with the focus of comparative psychologists on behaviour in artificial laboratory situations such as mazes and puzzle boxes. Ethologists view the naturalistic approach as crucial because it reveals the environmental and social circumstances in which the behaviour originally evolved, and prepares the way for more realistically designed laboratory experiments. The approach goes back to the stress that Charles Darwin placed on hereditary contributions to behaviour in all species, including humans. Viewing behaviour as a product of evolutionary history has helped to elucidate many otherwise puzzling aspects of its biology and has paved the way for the new science of neuroethology, concerned with how the structure and functioning of the brain controls behaviour and makes learning possible.

A central concept in classical ethology is that of the innate release mechanism. If a species has had a long history of experience with certain stimuli, especially those involving survival and reproduction, then to the extent that genes affect the ability to attend closely to such stimuli, natural selection leads to adaptations enhancing responsiveness to them. A common first step in the study of these adaptations was investigation of the development of responsiveness to such stimuli in infancy, focusing on situations that the ethologist knew to be especially relevant to survival. The term innate releasing mechanism, set forth by Tinbergen and Lorenz, eschews notions of innate mental imagery and has proved fertile in understanding how genes influence behavioural development, and in focusing attention of neuroethologists on inborn physiological mechanisms that permit learning while encouraging the infant to attend closely to very specific stimuli, the nature of which varies from species to species according to differences in ecology and social organization.

Fig. Ethology

In birds, such as the herring gull, innate release mechanisms are also better thought of as having evolved to guide processes of perceptual learning, rather than to design animals as though they were automata. Learning is as important in the development of the behaviour of many animals as in the human species. Yet as the young organism interacts with social and physical environments with which it has evolved adaptive relationships, the course that learning takes in nature is guided and profoundly influenced by the innate predispositions that the organism brings to bear on dealing with the situation.

Perceptions of the external world provide a basis for both thought and action. It is a fundamental axiom of ethology that each organism's brain is armed with genetically determined programmes of action which, in their own way, are as predictable and controlled as the genetic programmes for developing anatomical structures such as a brain or a face. Ethologists have shown that it is possible to reconcile the need to modify patterns of action on the basis of experience with the possession of basic patterns of action that are coordinated by the brain, innately controlled, and often distinct from species to species. The concept of the fixed action pattern remains fundamental to understanding the development of the ability to act. Innate "motor programmes," generated by the brain, are the natural units out of which behaviour emerges during development. Each of these programmes designates not a single completely stereotyped action, but a range of options which are limited but sufficient that selection among them allows for adjustments through experience. Close study reveals that sometimes the modifiability of actions lies in the potential flexibility of orientation, timing, and sequencing of actions rather than in the basic patterns or coordinations from which complex actions are built up.

Thus nervous systems make some behavioural adjustments promptly and easily and others only with much greater difficulty, in harmony with species differences in the requirements for patterns of action as dictated by the species' structure and mode of life. Ethologists have found repeatedly that while social experience is vital in many animals for normal development of actions and responses, animals reared in restricted environments may still develop many units of action that are normal. The animal has to learn, however, how to put them together in an adaptive sequence.

Modern research on the ethology of learning began when Lorenz discovered imprinting in geese. He found that if he led a flock of newly hatched goslings himself they became imprinted on him. When mature, they would court people as though confused about their own species identity. Learning occurred very rapidly and tended to be restricted to a short sensitive phase early in life. The learning is highly focused by genetically determined preferences both to follow a parent-object with particular appearance and emitting species-specific calls, and also to learn most quickly and accurately at a particular stage of development. The interplay between nature (genetic predisposition) and nurture (environmental influence) in learning is displayed especially clearly in imprinting, hence its special interest to biologists and psychiatrists. Indications are that it is not concerned so much with learning about species as with learning to recognize individual parents and kin, both to ensure mating with one's own kind and to avoid incestuous inbreeding.

There are many forms of imprinting. So-called filial imprinting, ensuring that ducklings and goslings follow only their parent, is distinct from sexual imprinting, affecting mate choice in adulthood; the sensitive phases for learning are different in each case. Imprinting-like processes also shape the development of food preferences and abilities to use the Sun and stars in navigation.

Unlike psychological studies of animal learning in the laboratory, which have tended to favour the "blank-slate" view of the brain's contribution to learning, ethology emphasizes the need to understand all aspects of the biology of a species under study before one can hope to understand how the animal learns to cope with the many complexities of individual existence and social living. Thus ethology may lead not only to an understanding of how natural behaviour evolves, but also to new insights into how brains help organisms learn to cope with social and environmental problems confronting them as individuals.

Ethology is the zoological study of animal behaviour. Ethologists have a special interest in genetically-programmed behaviours known as instincts. The predictable behavioural programmes are inherited by animals through their parents and portions of the programmes are open to natural selection and modification. Thus, these behaviours are phylogenetic adaptations that have an evolutionary history. This often leads to a comparative approach and has led researchers to search for the biological basis of human behaviour by comparing our activities to those of our close relative (other primates; especially chimps).

There are two schools of thought as to how animals acquire their behaviour patterns. Some hold to the view that animals, including humans, learn all their behaviour during the course of ontogenetic development. Ducks, for example, learn how to quack like a duck and don't honk like a goose because they hear their parents while within the egg). Experiments have since shown that these behaviours are built-in and not learned.

Very complex behaviour patterns can be passed on through the genes. A spider's orb web, for example, is built perfectly the first time a spider attempts construction, despite the fact that they may have no prior experience with webs (Fig 1). Female spiders construct egg sacs in the Fall, and then die. The spiderlings emerge in the spring, and never had experience with an orb web before building their own.

Ethology differs from the study of Animal Behaviour, in that animals behaviourists generally are interested in learned behaviours while ethologists concentrate on innate behaviours. Also, animals behaviourists tend to be trained in psychology, while ethologists are zoologists. Animal behaviourists tend to work with "bright" animals, such as rats; putting them through trials with mazes and Skinner boxes. While the study of learned behaviour is both important and immediately applicable to human psychology, these behaviours cannot have an evolutionary basis.

Konrad Lorenz, along with Karl von Frisch and Nikolaas Tinbergen are generally recognized as the "fathers of ethology". But the origins of ethology can be traced back to Charles Darwin and his work on the expressive movements of man and animal). Darwin was the first to use a comparative phylogenetic method in the study of behaviour.

No discussion of ethology would be complete without mentioning Irenäus Eibl-Eibesfeldt, who was the first to successfully apply ethological methods to the study of human behaviour. He carefully recorded the activities of humans using a side-viewing camera so the subjects didn't know they were being observed. By comparing gestures and body language across cultures, he identified numerous innate behaviour patterns in humans.

Much of the work in ethology revolves around problems in animal communication. In fact, this is the work for which von Frisch is best known. He teased apart the "dance language" of bees (Fig 4). Worker bees dance to communicate the distance, direction, and quality of a feeding place they have found while foraging around the hive. The dance is performed on the vertical comb of the hive. If the feeding place is directly toward the sun, the bees dance up. If the feeding place is directly away from the sun, they will dance down. Angles to the sun are communicated. As the sun moves across the sky, the bees adjust their dance so that the direction is always up-to-date. If, for example, a bee finds food at sunrise and indicates that it is located directly toward the sun, by the time the sun sets, the bee will be dancing down. Thus, bees have an internal biological clock that tells them where the sun should be in the sky without them having to go out and check.

Distance to the feeding place is communicated by the number of waggles at the centre part of the dance. More waggles indicate greater distance. Sites with lots of high quality food are communicated by the vigor of the dance (very good sites produce frantic dancing). In addition, other workers paying attention to the dancer touch their sister and can determine what type of flower she visited by the pollen stuck on her hairs. A second dance, a round dance, is used when the food source is very close to the hive.

The type of communication an animal uses can be described by the sensory modality used to receive the signal. Examples can be seen in the following table:

Sensory Modality	Example
Chemoreception	Pheromones (group cohesion, territory marking, individual recognition, sex pheromones, etc.)
Mechanoreception	Vibrational cues including substrate-coupled (spider on web) and sound (bird and cricket song)
Visual	Bird courtship dances, aggressive stances of canids, firefly light patterns.
Electrical	Electric eels and weakly electric fish communicate with one another through charged pulses.
Psychic	That's how those voices in your head get there!

EVOLUTIONARY BIOLOGY

Evolution is the cornerstone of modern biology. It unites all the fields of biology under one theoretical umbrella. It is not a difficult concept, but very few people — the majority of biologists included — have a satisfactory grasp of it. One common mistake is believing that species can be arranged on an evolutionary ladder from bacteria through "lower" animals, to "higher" animals and, finally, up to man. Mistakes permeate popular science expositions of evolutionary biology. Mistakes even filter into biology journals and texts. For example, Lodish, et. al., in their cell biology text, proclaim, "It was Charles Darwin's great insight that organisms are all related in a great chain of being..." In fact, the idea of a great chain of being, which traces to Linnaeus, was overturned by Darwin's idea of common descent.

Misunderstandings about evolution are damaging to the study of evolution and biology as a whole. People who have a general interest in science are likely to dismiss evolution as a soft science after absorbing the pop science nonsense that abounds. The impression of it being a soft science is reinforced when biologists in unrelated fields speculate publicly about evolution.

This is a brief introduction to evolutionary biology. I attempt to explain basics of the theory of evolution and correct many of the misconceptions.

WHAT IS EVOLUTION?

Evolution is a change in the gene pool of a population over time. A gene is a hereditary unit that can be passed on unaltered for many generations. The gene pool is the set of all genes in a species or population.

The English moth, Biston betularia, is a frequently cited example of observed evolution. [evolution: a change in the gene pool] In this moth there are two colour morphs, light and dark. H. B. D. Kettlewell found that dark moths constituted less than 2% of the population prior to 1848. The frequency of the dark morph increased in the years following. By 1898, the 95% of the moths in Manchester and other highly industrialized areas were of the dark type. Their frequency was less in rural areas. The moth population changed from mostly light coloured moths to mostly dark coloured moths. The moths' colour was primarily determined by a single gene. [gene: a hereditary unit] So, the change in frequency of dark coloured moths represented a change in the gene pool. [gene pool: the set all of genes in a population] This change was, by definition, evolution.

The increase in relative abundance of the dark type was due to natural selection. The late eighteen hundreds was the time of England's industrial revolution. Soot from factories darkened the birch trees the moths landed on. Against a sooty background, birds could see the lighter coloured moths better and ate more of them. As a result, more dark moths survived until reproductive age and left offspring. The greater number of offspring left by dark moths is what caused their increase in frequency. This is an example of natural selection.

Populations evolve. [evolution: a change in the gene pool] In order to understand evolution, it is necessary to view populations as a collection of individuals, each harboring a different set of traits. A single organism is never typical of an entire population unless there is no variation within that population. Individual organisms do not evolve, they retain the same genes throughout their life. When a population is evolving, the ratio of different genetic types is changing – each individual organism within a population does not change. For example, in the previous example, the frequency of black moths increased; the moths did not turn from light to gray to dark in concert. The process of evolution can be summarized in three sentences: Genes mutate. [gene: a hereditary unit] Individuals are selected. Populations evolve.

Evolution can be divided into microevolution and macroevolution. The kind of evolution documented above is microevolution. Larger changes, such as when a new species is formed, are called macroevolution. Some biologists feel the mechanisms of macroevolution are different from those of microevolutionary change. Others think the distinction between the two is arbitrary – macroevolution is cumulative microevolution.

The word evolution has a variety of meanings. The fact that all organisms are linked via descent to a common ancestor is often called evolution. The theory of how the first living organisms appeared is often called evolution. This should be called abiogenesis. And frequently, people use the word evolution when they really mean natural selection – one of the many mechanisms of evolution.

COMMON MISCONCEPTIONS ABOUT EVOLUTION

Evolution can occur without morphological change; and morphological change can occur without evolution. Humans are larger now than in the recent past, a result of better diet and medicine. Phenotypic changes, like this, induced solely by changes in environment do not count as evolution because they are not heritable; in other words the change is not passed on to the organism's offspring. Phenotype is the morphological, physiological, biochemical, behavioural and other properties exhibited by a living organism. An organism's phenotype is determined by its genes and its environment. Most changes due to environment are fairly subtle, for example size differences. Large scale phenotypic changes are obviously due to genetic changes, and therefore are evolution.

Evolution is not progress. Populations simply adapt to their current surroundings. They do not necessarily become better in any absolute sense over time. A trait or strategy that is successful at one time may be unsuccessful at another. Paquin and Adams demonstrated this experimentally. They founded a yeast culture and maintained it for many generations. Occasionally, a mutation would arise that allowed its bearer to reproduce better than its contemporaries. These mutant strains would crowd out the formerly dominant strains. Samples of the most successful strains from the culture were taken at a variety of times. In later competition experiments, each strain would outcompete the immediately previously dominant type in a culture. However, some earlier isolates could outcompete strains that arose late in the experiment. Competitive ability of a strain was always better than its previous type, but competitiveness in a general sense was not increasing. Any organism's success depends on the behaviour of its contemporaries. For most traits or behaviours there is likely no optimal design or strategy, only contingent ones. Evolution can be like a game of paper/scissors/rock.

Organisms are not passive targets of their environment. Each species modifies its own environment. At the least, organisms remove nutrients from and add waste to their surroundings. Often, waste products benefit other species. Animal dung is fertilizer for plants. Conversely, the oxygen we breathe is a waste product of plants. Species do not simply change to fit their environment; they modify their environment to suit them as well. Beavers build a dam to create a pond suitable to sustain them and raise young. Alternately, when the environment changes, species can migrate to suitable climes or seek out microenvironments to which they are adapted.

GENETIC VARIATION

Evolution requires genetic variation. If there were no dark moths, the population could not have evolved from mostly light to mostly dark. In order for continuing evolution there must be mechanisms to increase or create genetic variation and mechanisms to decrease it. Mutation is a change in a gene. These changes are the source of new genetic variation. Natural selection operates on this variation.

Genetic variation has two components: allelic diversity and non- random associations of alleles. Alleles are different versions of the same gene. For example, humans can have A, B or O alleles that determine one aspect of their blood type. Most animals, including humans, are diploid – they contain two alleles for every gene at every locus, one inherited from their mother and one inherited from their father. Locus is the location of a gene on a chromosome.

Humans can be AA, AB, AO, BB, BO or OO at the blood group locus. If the two alleles at a locus are the same type (for instance two A alleles) the individual would be called homozygous. An individual with two different alleles at a locus (for example, an AB individual) is called heterozygous. At any locus there can be many different alleles in a population, more alleles than any single organism can possess. For example, no single human can have an A, B and an O allele.

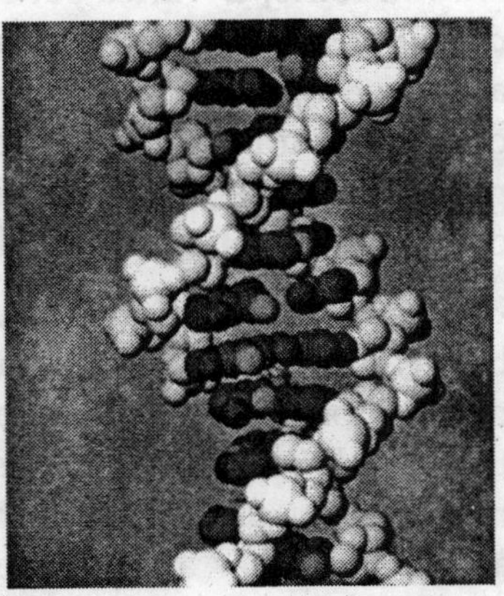

Fig. Genetic Variation

Considerable variation is present in natural populations. At 45 percent of loci in plants there is more than one allele in the gene pool. [allele: alternate version of a gene (created by mutation)] Any given plant is likely to be heterozygous at about 15 percent of its loci. Levels of genetic variation in animals range from roughly 15% of loci having more than one allele (polymorphic) in birds, to over 50% of loci being polymorphic in insects. Mammals and reptiles are polymorphic at about 20% of their loci - - amphibians and fish are polymorphic at around 30% of their loci. In most populations, there are enough loci and enough different alleles that every individual, identical twins excepted, has a unique combination of alleles.

Linkage disequilibrium is a measure of association between alleles of two different genes. [allele: alternate version of a gene] If two alleles were found together in organisms more often than would be expected, the alleles are in linkage disequilibrium. If there two loci in an organism (A and B) and two alleles at each of these loci (A1, A2, B1 and B2) linkage disequilibrium (D) is calculated as D = f(A1B1) * f(A2B2) - f(A1B2) * f(A2B1) (where f(X) is the frequency of X in the population). [Loci (plural of locus): location of a gene on a chromosome] D varies between -1/4 and 1/4; the greater the deviation from zero, the greater the linkage. The sign is simply a consequence of how the alleles are numbered. Linkage disequilibrium can be the result of physical proximity of the genes. Or, it can be maintained by natural selection if some combinations of alleles work better as a team.

Natural selection maintains the linkage disequilibrium between colour and pattern alleles in Papilio memnon. [linkage disequilibrium: association between alleles at different loci] In this moth species, there is a gene that determines wing morphology. One allele at this locus leads to a moth that has a tail; the other allele codes for a untailed moth. There is another gene that determines if the wing is brightly or darkly coloured. There are thus four possible types of moths: brightly coloured moths with and without tails, and dark moths with and without tails. All four can be produced when moths are brought into the lab and bred. However, only two of these types of moths are found in the wild: brightly coloured moths with tails and darkly coloured moths without tails. The non-random association is maintained by natural selection. Bright, tailed moths mimic the pattern of an unpalatable species. The dark morph is cryptic. The other two combinations are neither mimetic nor cryptic and are quickly eaten by birds.

Assortative mating causes a non-random distribution of alleles at a single locus. [locus: location of a gene on a chromosome] If there are two alleles (A and a) at a locus with frequencies p and q, the frequency of the three possible genotypes (AA, Aa and aa) will be p^2, 2pq and q^2, respectively. For example, if the frequency of A is 0.9 and the frequency of a is 0.1, the frequencies of AA, Aa and aa individuals are: 0.81, 0.18 and 0.01. This distribution is called the Hardy-Weinberg equilibrium.

Non-random mating results in a deviation from the Hardy-Weinberg distribution. Humans mate assortatively according to race; we are more likely to mate with someone of own race than another. In populations that mate this way, fewer heterozygotes are found than would be predicted under random mating. [heterozygote: an organism that has two different alleles at a locus] A decrease in heterozygotes can be the result of mate choice, or simply the result of population subdivision. Most organisms have a limited dispersal capability, so their mate will be chosen from the local population.

EVOLUTION WITHIN A LINEAGE

In order for continuing evolution there must be mechanisms to increase or create genetic variation and mechanisms to decrease it. The mechanisms of evolution are mutation, natural selection, genetic drift, recombination and gene flow. We have grouped them into two classes — those that decrease genetic variation and those that increase it.

Mechanisms That Decrease Genetic Variation

Natural Selection

Some types of organisms within a population leave more offspring than others. Over time, the frequency of the more prolific type will increase. The difference in reproductive capability is called natural selection. Natural selection is the only mechanism of adaptive evolution; it is defined as differential reproductive success of pre- existing classes of genetic variants in the gene pool.

The most common action of natural selection is to remove unfit variants as they arise via mutation. [natural selection: differential reproductive success of genotypes] In other words, natural selection usually prevents new alleles from increasing in frequency. This led a famous evolutionist, George Williams, to say "Evolution proceeds in spite of natural selection."

Natural selection can maintain or deplete genetic variation depending on how it acts. When selection acts to weed out deleterious alleles, or causes an allele to sweep to fixation, it depletes genetic variation. When heterozygotes are more fit than either of the homozygotes, however, selection causes genetic variation to be maintained. Heterozygote: an organism that has two different alleles at a locus. Homozygote, an organism that has two identical alleles at a locus. This is called balancing selection. An example of this is the maintenance of sickle-cell alleles in human populations subject to malaria. Variation at a single locus determines whether red blood cells are shaped normally or sickled. If a human has two alleles for sickle-cell, he/she develops anemia — the shape of sickle-cells precludes them carrying normal levels of oxygen. However, heterozygotes who have one copy of the sickle-cell allele, coupled with one normal allele enjoy some resistance to malaria — the shape of sickled cells make it harder for the plasmodia (malaria causing agents) to enter the cell. Thus, individuals homozygous for the normal allele suffer more malaria than heterozygotes. Individuals homozygous for the sickle- cell are anemic. Heterozygotes have the highest fitness of these three types. Heterozygotes pass on both sickle-cell and normal alleles to the next generation. Thus, neither allele can be eliminated from the gene pool. The sickle-cell allele is at its highest frequency in regions of Africa where malaria is most pervasive.

Balancing selection is rare in natural populations (balancing selection: selection favoring heterozygotes). Only a handful of other cases beside the sickle-cell example have been found. At one time population geneticists thought balancing selection could be a general explanation for the levels of genetic variation found in natural populations. That is no longer the case. Balancing selection is only rarely found in natural populations. And, there are theoretical reasons why natural selection cannot maintain polymorphisms at several loci via balancing selection.

Individuals are selected. The example given earlier was an example of evolution via natural selection. Natural selection: differential reproductive success of genotypes: Dark coloured moths had a higher reproductive success because light coloured moths suffered a higher predation rate. The decline of light coloured alleles was caused by light coloured individuals being removed from the gene pool (selected against). Individual organisms either reproduce or fail to reproduce and are hence the unit of selection. One way alleles can change in frequency is to be housed in organisms with different reproductive rates. Genes are not the unit of selection (because their success depends on the organism's other genes as well); neither are groups of organisms a unit of selection. There are some exceptions to this "rule," but it is a good generalization.

Organisms do not perform any behaviours that are for the good of their species. An individual organism competes primarily with others of it own species for its reproductive success. Natural selection favours selfish behaviour because any truly altruistic act increases the recipient's reproductive success while lowering the donors. Altruists would disappear from a population as the non- altruists would reap the benefits, but not pay the costs, of altruistic acts. Many behaviours appear altruistic. Biologists, however, can demonstrate that these behaviours are only apparently altruistic. Cooperating with or helping other organisms is often the most selfish strategy for an animal. This is called reciprocal altruism. A good example of this is blood sharing in vampire bats. In these bats, those lucky enough to find a meal will often share part of it with an unsuccessful bat by regurgitating some blood into the other's mouth. Biologists have found that these bats form bonds with partners and help each other out when the other is needy. If a bat is found to be a "cheater," (he accepts blood when starving, but does not donate when his partner is) his partner will abandon him. The bats are thus not helping each other altruistically; they form pacts that are mutually beneficial.

Helping closely related organisms can appear altruistic; but this is also a selfish behaviour. Reproductive success (fitness) has two components; direct fitness and indirect fitness. Direct fitness is a measure of how many alleles, on average, a genotype contributes to the subsequent generation's gene pool by reproducing. Indirect fitness is a measure of how many alleles identical to its own it helps to enter the gene pool. Direct fitness plus indirect fitness is inclusive fitness. J. B. S. Haldane once remarked he would gladly drown, if by doing so he saved two siblings or eight cousins. Each of his siblings would share one half his alleles; his cousins, one eighth. They could potentially add as many of his alleles to the gene pool as he could.

Natural selection favours traits or behaviours that increase a genotype's inclusive fitness. Closely related organisms share many of the same alleles. In diploid species, siblings share on average at least 50% of their alleles. The percentage is higher if the parents are related. So, helping close relatives to reproduce gets an organism's own alleles better represented in the gene pool. The benefit of helping relatives increases dramatically in highly inbred species. In some cases, organisms will completely forgo reproducing and only help their relatives reproduce. Ants, and other eusocial insects, have sterile castes that only serve the queen and assist her reproductive efforts. The sterile workers are reproducing by proxy.

The words selfish and altruistic have connotations in everyday use that biologists do not intend. Selfish simply means behaving in such a way that one's own inclusive fitness is maximized; altruistic means behaving in such a way that another's fitness is increased at the expense of ones' own. Use of the words selfish and altruistic is not meant to imply that organisms consciously understand their motives.

The opportunity for natural selection to operate does not induce genetic variation to appear – selection only distinguishes between existing variants. Variation is not possible along every imaginable axis, so all possible adaptive solutions are not open to populations. To pick a somewhat ridiculous example, a steel shelled turtle might be an improvement over regular turtles. Turtles are killed quite a bit by cars these days because when confronted with danger, they retreat into their shells – this is not a great strategy against a two ton automobile. However, there is no variation in metal content of shells, so it would not be possible to select for a steel shelled turtle.

Here is a second example of natural selection. Geospiza fortis lives on the Galapagos islands along with fourteen other finch species. It feeds on the seeds of the plant Tribulus cistoides, specializing on the smaller seeds. Another species, G. Magnirostris, has a larger beak and specializes on the larger seeds. The health of these bird populations depends on seed production. Seed production, in turn, depends on the arrival of wet season. In 1977, there was a drought. Rainfall was well below normal and fewer seeds were produced. As the season progressed, the G. fortis population depleted the supply of small seeds. Eventually, only larger seeds remained. Most of the finches starved; the population plummeted from about twelve hundred birds to less than two hundred. Peter Grant, who had been studying these finches, noted that larger beaked birds fared better than smaller beaked ones. These larger birds had offspring with correspondingly large beaks. Thus, there was an increase in the proportion of large beaked birds in the population the next generation. To prove that the change in bill size in Geospiza fortis was an evolutionary change, Grant had to show that differences in bill size were at least partially genetically based. He did so by crossing finches of various beak sizes and showing that a finch's beak size was influenced by its parent's genes. Large beaked birds had large beaked offspring; beak size was not due to environmental differences (in parental care, for example).

Natural selection may not lead a population to have the optimal set of traits. In any population, there would be a certain combination of possible alleles that would produce the optimal set of traits (the global optimum); but there are other sets of alleles that would yield a population almost as adapted (local optima). Transition from a local optimum to the global optimum may be hindered or forbidden because the population would have to pass through less adaptive states to make the transition. Natural selection only works to bring populations to the nearest optimal point. This idea is Sewall Wright's adaptive landscape. This is one of the most influential models that shape how evolutionary biologists view evolution.

Natural selection does not have any foresight. It only allows organisms to adapt to their current environment. Structures or behaviours do not evolve for future utility. An organism adapts to its environment at each stage of its evolution. As the environment changes, new traits may be selected for. Large changes in populations are the result of cumulative natural selection. Changes are introduced into the population by mutation; the small minority of these changes that result in a greater reproductive output of their bearers are amplified in frequency by selection.

Complex traits must evolve through viable intermediates. For many traits, it initially seems unlikely that intermediates would be viable. What good is half a wing? Half a wing may be no good for flying, but it may be useful in other ways. Feathers are thought to have evolved as insulation (ever worn a down jacket?) and/or as a way to trap insects. Later, proto-birds may have learned to glide when leaping from tree to tree. Eventually, the feathers that originally served as insulation now became co-opted for use in flight. A trait's current utility is not always indicative of its past utility. It can evolve for one purpose, and be used later for another. A trait evolved for its current utility is an adaptation; one that evolved for another utility is an exaptation. An example of an exaptation is a penguin's wing. Penguins evolved from flying ancestors; now they are flightless and use their wings for swimming.

COMMON MISCONCEPTIONS ABOUT SELECTION

Selection is not a force in the sense that gravity or the strong nuclear force is. However, for the sake of brevity, biologists sometimes refer to it that way. This often leads to some confusion when biologists speak of selection "pressures." This implies that the environment "pushes" a population to more adapted state. This is not the case. Selection merely favours beneficial genetic changes when they occur by chance – it does not contribute to their appearance. The potential for selection to act may long precede the appearance of selectable genetic variation. When selection is spoken of as a force, it often seems that it is has a mind of its own; or as if it was nature personified. This most often occurs when biologists are waxing poetic about selection. This has no place in scientific discussions of evolution. Selection is not a guided or cognizant entity; it is simply an effect.

A related pitfall in discussing selection is anthropomorphizing on behalf of living things. Often conscious motives are seemingly imputed to organisms, or even genes, when discussing evolution. This happens most frequently when discussing animal behaviour. Animals are often said to perform some behaviour because selection will favour it. This could more accurately worded as "animals that, due to their genetic composition, perform this behaviour tend to be favored by natural selection relative to those who, due to their genetic composition, don't." Such wording is cumbersome. To avoid this, biologists often anthropomorphize. This is unfortunate because it often makes evolutionary arguments sound silly. Keep in mind this is only for convenience of expression.

The phrase "survival of the fittest" is often used synonymously with natural selection. The phrase is both incomplete and misleading. For one thing, survival is only one component of selection – and perhaps one of the less important ones in many populations. For example, in polygynous species, a number of males survive to reproductive age, but only a few ever mate. Males may differ little in their ability to survive, but greatly in their ability to attract mates – the difference in reproductive success stems mainly from the latter consideration. Also, the word fit is often confused with physically fit. Fitness, in an evolutionary sense, is the average reproductive output of a class of genetic variants in a gene pool. Fit does not necessarily mean biggest, fastest or strongest.

SEXUAL SELECTION

In many species, males develop prominent secondary sexual characteristics. A few oft cited examples are the peacock's tail, colouring and patterns in male birds in general, voice calls in frogs and flashes in fireflies. Many of these traits are a liability from the standpoint of survival. Any ostentatious trait or noisy, attention getting behaviour will alert predators as well as potential mates. How then could natural selection favour these traits?

Natural selection can be broken down into many components, of which survival is only one. Sexual attractiveness is a very important component of selection, so much so that biologists use the term sexual selection when they talk about this subset of natural selection.

Sexual selection is natural selection operating on factors that contribute to an organism's mating success. Traits that are a liability to survival can evolve when the sexual attractiveness of a trait outweighs the liability incurred for survival. A male who lives a short time, but produces many offspring is much more successful than a long lived one that produces few. The former's genes will eventually dominate the gene pool of his species. In many species, especially polygynous species where only a few males monopolize all the females, sexual selection has caused pronounced sexual dimorphism. In these species males compete against other males for mates. The competition can be either direct or mediated by female choice. In species where females choose, males compete by displaying striking phenotypic characteristics and/or performing elaborate courtship behaviours. The females then mate with the males that most interest them, usually the ones with the most outlandish displays. There are many competing theories as to why females are attracted to these displays.

The good genes model states that the display indicates some component of male fitness. A good genes advocate would say that bright colouring in male birds indicates a lack of parasites. The females are cueing on some signal that is correlated with some other component of viability.

Selection for good genes can be seen in sticklebacks. In these fish, males have red colouration on their sides. Milinski and Bakker showed that intensity of colour was correlated to both parasite load and sexual attractiveness. Females preferred redder males. The redness indicated that he was carrying fewer parasites.

Evolution can get stuck in a positive feedback loop. Another model to explain secondary sexual characteristics is called the runaway sexual selection model. R. A. Fisher proposed that females may have an innate preference for some male trait before it appears in a population. Females would then mate with male carriers when the trait appears. The offspring of these matings have the genes for both the trait and the preference for the trait. As a result, the process snowballs until natural selection brings it into check. Suppose that female birds prefer males with longer than average tail feathers. Mutant males with longer than average feathers will produce more offspring than the short feathered males. In the next generation, average tail length will increase. As the generations progress, feather length will increase because females do not prefer a specific length tail, but a longer than average tail. Eventually tail length will increase to the point were the liability to survival is matched by the sexual attractiveness of the trait and an equilibrium will be established. Note that in many exotic birds male plumage is often very showy and many species do in fact have males with greatly elongated feathers. In some cases these feathers are shed after the breeding season.

None of the above models are mutually exclusive. There are millions of sexually dimorphic species on this planet and the forms of sexual selection probably vary amongst them.

GENETIC DRIFT

Allele frequencies can change due to chance alone. This is called genetic drift. Drift is a binomial sampling error of the gene pool. What this means is, the alleles that form the next generation's gene pool are a sample of the alleles from the current generation. When sampled from a population, the frequency of alleles differs slightly due to chance alone.

Alleles can increase or decrease in frequency due to drift. The average expected change in allele frequency is zero, since increasing or decreasing in frequency is equally probable. A small percentage of alleles may continually change frequency in a single direction for several generations just as flipping a fair coin may, on occasion, result in a string of heads or tails. A very few new mutant alleles can drift to fixation in this manner.

In small populations, the variance in the rate of change of allele frequencies is greater than in large populations. However, the overall rate of genetic drift (measured in substitutions per generation) is independent of population size. [genetic drift: a random change in allele frequencies] If the mutation rate is constant, large and small populations lose alleles to drift at the same rate. This is because large populations will have more alleles in the gene pool, but they will lose them more slowly. Smaller populations will have fewer alleles, but these will quickly cycle through. This assumes that mutation is constantly adding new alleles to the gene pool and selection is not operating on any of these alleles.

Sharp drops in population size can change allele frequencies substantially. When a population crashes, the alleles in the surviving sample may not be representative of the precrash gene pool. This change in the gene pool is called the founder effect, because small populations of organisms that invade a new territory (founders) are subject to this. Many biologists feel the genetic changes brought about by founder effects may contribute to isolated populations developing reproductive isolation from their parent populations. In sufficiently small populations, genetic drift can counteract selection. [genetic drift: a random change in allele frequencies] Mildly deleterious alleles may drift to fixation.

Wright and Fisher disagreed on the importance of drift. Fisher thought populations were sufficiently large that drift could be neglected. Wright argued that populations were often divided into smaller subpopulations.

Drift could cause allele frequency differences between subpopulations if gene flow was small enough. If a subpopulation was small enough, the population could even drift through fitness valleys in the adaptive landscape. Then, the subpopulation could climb a larger fitness hill. Gene flow out of this subpopulation could contribute to the population as a whole adapting. This is Wright's Shifting Balance theory of evolution.

Both natural selection and genetic drift decrease genetic variation. If they were the only mechanisms of evolution, populations would eventually become homogeneous and further evolution would be impossible. There are, however, mechanisms that replace variation depleted by selection and drift.

MECHANISMS THAT INCREASE GENETIC VARIATION

MUTATION

The cellular machinery that copies DNA sometimes makes mistakes. These mistakes alter the sequence of a gene. This is called a mutation. There are many kinds of mutations. A point mutation is a mutation in which one "letter" of the genetic code is changed to another. Lengths of DNA can also be deleted or inserted in a gene; these are also mutations. Finally, genes or parts of genes can become inverted or duplicated. Typical rates of mutation are between 10^{-10} and 10^{-12} mutations per base pair of DNA per generation.

Most mutations are thought to be neutral with regards to fitness. (Kimura defines neutral as $|s| < 1/2Ne$, where s is the selective coefficient and Ne is the effective population size.) Only a small portion of the genome of eukaryotes contains coding segments. And, although some non-coding DNA is involved in gene regulation or other cellular functions, it is probable that most base changes would have no fitness consequence.

Most mutations that have any phenotypic effect are deleterious. Mutations that result in amino acid substitutions can change the shape of a protein, potentially changing or eliminating its function. This can lead to inadequacies in biochemical pathways or interfere with the process of development. Organisms are sufficiently integrated that most random changes will not produce a fitness benefit. Only a very small percentage of mutations are beneficial. The ratio of neutral to deleterious to beneficial mutations is unknown and probably varies with respect to details of the locus in question and environment.

Mutation limits the rate of evolution. The rate of evolution can be expressed in terms of nucleotide substitutions in a lineage per generation. Substitution is the replacement of an allele by another in a population. This is a two step process: First a mutation occurs in an individual, creating a new allele.

This allele subsequently increases in frequency to fixation in the population. The rate of evolution is k = 2Nvu (in diploids) where k is nucleotide substitutions, N is the effective population size, v is the rate of mutation and u is the proportion of mutants that eventually fix in the population.

Mutation need not be limiting over short time spans. The rate of evolution expressed above is given as a steady state equation; it assumes the system is at equilibrium. Given the time frames for a single mutant to fix, it is unclear if populations are ever at equilibrium. A change in environment can cause previously neutral alleles to have selective values; in the short term evolution can run on "stored" variation and thus is independent of mutation rate. Other mechanisms can also contribute selectable variation. Recombination creates new combinations of alleles (or new alleles) by joining sequences with separate microevolutionary histories within a population. Gene flow can also supply the gene pool with variants. Of course, the ultimate source of these variants is mutation.

The Fate of Mutant Alleles

Mutation creates new alleles. Each new allele enters the gene pool as a single copy amongst many. Most are lost from the gene pool, the organism carrying them fails to reproduce, or reproduces but does not pass on that particular allele. A mutant's fate is shared with the genetic background it appears in. A new allele will initially be linked to other loci in its genetic background, even loci on other chromosomes. If the allele increases in frequency in the population, initially it will be paired with other alleles at that locus — the new allele will primarily be carried in individuals heterozygous for that locus. The chance of it being paired with itself is low until it reaches intermediate frequency. If the allele is recessive, its effect won't be seen in any individual until a homozygote is formed. The eventual fate of the allele depends on whether it is neutral, deleterious or beneficial.

Neutral Alleles

Most neutral alleles are lost soon after they appear. The average time (in generations) until loss of a neutral allele is

2(Ne/N) ln(2N) where N is the effective population size (the number of individuals contributing to the next generation's gene pool) and N is the total population size. Only a small percentage of alleles fix. Fixation is the process of an allele increasing to a frequency at or near one. The probability of a neutral allele fixing in a population is equal to its frequency. For a new mutant in a diploid population, this frequency is 1/2N.

If mutations are neutral with respect to fitness, the rate of substitution (k) is equal to the rate of mutation(v). This does not mean every new mutant eventually reaches fixation. Alleles are added to the gene pool by mutation at the same rate they are lost to drift. For neutral alleles that do fix, it takes an average of 4N generations to do so. However, at equilibrium there are multiple alleles segregating in the population. In small populations, few mutations appear each generation. The ones that fix do so quickly relative to large populations. In large populations, more mutants appear over the generations. But, the ones that fix take much longer to do so. Thus, the rate of neutral evolution (in substitutions per generation) is independent of population size.

The rate of mutation determines the level of heterozygosity at a locus according to the neutral theory. Heterozygosity is simply the proportion of the population that is heterozygous. Equilibrium heterozygosity is given as H = 4Nv/[4Nv+1] (for diploid populations). H can vary from a very small number to almost one. In small populations, H is small (because the equation is approximately a very small number divided by one). In (biologically unrealistically) large populations, heterozygosity approaches one (because the equation is approximately a large number divided by itself). Directly testing this model is difficult because N and v can only be estimated for most natural populations. But, heterozygosities are believed to be too low to be described by a strictly neutral model. Solutions offered by neutralists for this discrepancy include hypothesizing that natural populations may not be at equilibrium.

At equilibrium there should be a few alleles at intermediate frequency and many at very low frequencies. This is the Ewens- Watterson distribution. New alleles enter a population every generation, most remain at low frequency until they are lost. A few drift to intermediate frequencies, a very few drift all the way to fixation. In Drosophila pseudoobscura, the protein Xanthine dehydrogenase (Xdh) has many variants. In a single population, Keith, et. al., found that 59 of 96 proteins were of one type, two others were represented ten and nine times and nine other types were present singly or in low numbers.

Deleterious Alleles

Deleterious mutants are selected against but remain at low frequency in the gene pool. In diploids, a deleterious recessive mutant may increase in frequency due to drift. Selection cannot see it when it is masked by a dominant allele. Many disease causing alleles remain at low frequency for this reason. People who are carriers do not suffer the negative effect of the allele. Unless they mate with another carrier, the allele may simply continue to be passed on. Deleterious alleles also remain in populations at a low frequency due to a balance between recurrent mutation and selection. This is called the mutation load.

Beneficial Alleles

Most new mutants are lost, even beneficial ones. Wright calculated that the probability of fixation of a beneficial allele is 2s. (This assumes a large population size, a small fitness benefit, and that heterozygotes have an intermediate fitness. A benefit of 2s yields an overall rate of evolution: k=4Nvs where v is the mutation rate to beneficial alleles) An allele that conferred a one percent increase in fitness only has a two percent chance of fixing. The probability of fixation of beneficial type of mutant is boosted by recurrent mutation. The beneficial mutant may be lost several times, but eventually it will arise and stick in a population. (Recall that even deleterious mutants recur in a population.)

Directional selection depletes genetic variation at the selected locus as the fitter allele sweeps to fixation. Sequences linked to the selected allele also increase in frequency due to hitchhiking. The lower the rate of recombination, the larger the window of sequence that hitchhikes. Begun and Aquadro compared the level of nucleotide polymorphism within and between species with the rate of recombination at a locus. Low levels of nucleotide polymorphism within species coincided with low rates of recombination. This could be explained by molecular mechanisms if recombination itself was mutagenic. In this case, recombination with also be correlated with nucleotide divergence between species. But, the level of sequence divergence did not correlate with the rate of recombination. Thus, they inferred that selection was the cause. The correlation between recombination and nucleotide polymorphism leaves the conclusion that selective sweeps occur often enough to leave an imprint on the level of genetic variation in natural populations.

One example of a beneficial mutation comes from the mosquito Culex pipiens. In this organism, a gene that was involved with breaking down organophosphates - common insecticide ingredients -became duplicated. Progeny of the organism with this mutation quickly swept across the worldwide mosquito population. There are numerous examples of insects developing resistance to chemicals, especially DDT which was once heavily used in this country. And, most importantly, even though "good" mutations happen much less frequently than "bad" ones, organisms with "good" mutations thrive while organisms with "bad" ones die out.

If beneficial mutants arise infrequently, the only fitness differences in a population will be due to new deleterious mutants and the deleterious recessives. Selection will simply be weeding out unfit variants. Only occasionally will a beneficial allele be sweeping through a population. The general lack of large fitness differences segregating in natural populations argues that beneficial mutants do indeed arise infrequently. However, the impact of a beneficial mutant on the level of variation at a locus can be large and lasting. It takes many generations for a locus to regain appreciable levels of heterozygosity following a selective sweep.

Recombination

Each chromosome in our sperm or egg cells is a mixture of genes from our mother and our father. Recombination can be thought of as gene shuffling. Most organisms have linear chromosomes and their genes lie at specific location (loci) along them. Bacteria have circular chromosomes. In most sexually reproducing organisms, there are two of each chromosome type in every cell. For instance in humans, every chromosome is paired, one inherited from the mother, the other inherited from the father. When an organism produces gametes, the gametes end up with only one of each chromosome per cell. Haploid gametes are produced from diploid cells by a process called meiosis.

In meiosis, homologous chromosomes line up. The DNA of the chromosome is broken on both chromosomes in several places and rejoined with the other strand. Later, the two homologous chromosomes are split into two separate cells that divide and become gametes. But, because of recombination, both of the chromosomes are a mix of alleles from the mother and father.

Recombination creates new combinations of alleles. Alleles that arose at different times and different places can be brought together. Recombination can occur not only between genes, but within genes as well. Recombination within a gene can form a new allele. Recombination is a mechanism of evolution because it adds new alleles and combinations of alleles to the gene pool.

GENE FLOW

New organisms may enter a population by migration from another population. If they mate within the population, they can bring new alleles to the local gene pool. This is called gene flow. In some closely related species, fertile hybrids can result from interspecific matings. These hybrids can vector genes from species to species.

Gene flow between more distantly related species occurs infrequently. This is called horizontal transfer. One interesting case of this involves genetic elements called P elements. Margaret Kidwell found that P elements were transferred from some species in the Drosophila willistoni group to Drosophila melanogaster. These two species of fruit flies are distantly related and hybrids do not form. Their ranges do, however, overlap. The P elements were vectored into D. melanogaster via a parasitic mite that targets both these species. This mite punctures the exoskeleton of the flies and feeds on the "juices". Material, including DNA, from one fly can be transferred to another when the mite feeds. Since P elements actively move in the genome (they are themselves parasites of DNA), one incorporated itself into the genome of a melanogaster fly and subsequently spread through the species. Laboratory stocks of melanogaster caught prior to the 1940's lack of P elements. All natural populations today harbor them.

OVERVIEW OF EVOLUTION WITHIN A LINEAGE

Evolution is a change in the gene pool of a population over time; it can occur due to several factors. Three mechanisms add new alleles to the gene pool: mutation, recombination and gene flow. Two mechanisms remove alleles, genetic drift and natural selection. Drift removes alleles randomly from the gene pool. Selection removes deleterious alleles from the gene pool. The amount of genetic variation found in a population is the balance between the actions of these mechanisms.

Natural selection can also increase the frequency of an allele. Selection that weeds out harmful alleles is called negative selection. Selection that increases the frequency of helpful alleles is called positive, or sometimes positive Darwinian, selection. A new allele can also drift to high frequency. But, since the change in frequency of an allele each generation is random, nobody speaks of positive or negative drift.

Except in rare cases of high gene flow, new alleles enter the gene pool as a single copy. Most new alleles added to the gene pool are lost almost immediately due to drift or selection; only a small percent ever reach a high frequency in the population. Even most moderately beneficial alleles are lost due to drift when they appear. But, a mutation can reappear numerous times.

The fate of any new allele depends a great deal on the organism it appears in. This allele will be linked to the other alleles near it for many generations. A mutant allele can increase in frequency simply because it is linked to a beneficial allele at a nearby locus. This can occur even if the mutant allele is deleterious, although it must not be so deleterious as to offset the benefit of the other allele. Likewise a potentially beneficial new allele can be eliminated from the gene pool because it was linked to deleterious alleles when it first arose. An allele "riding on the coat tails" of a beneficial allele is called a hitchhiker. Eventually, recombination will bring the two loci to linkage equilibrium. But, the more closely linked two alleles are, the longer the hitchhiking will last.

The effects of selection and drift are coupled. Drift is intensified as selection pressures increase. This is because increased selection (i.e. a greater difference in reproductive success among organisms in a population) reduces the effective population size, the number of individuals contributing alleles to the next generation.

Adaptation is brought about by cumulative natural selection, the repeated sifting of mutations by natural selection. Small changes, favored by selection, can be the stepping-stone to further changes. The summation of large numbers of these changes is macroevolution.

Chapter 10

Physiology

Physiology (pronounced "fizzy-aw-low-jee") is the study of life. Physiology helps us understand how the body works, from the smallest part (cells) all the way to the whole body. It helps us understand how different parts of the body work together. For example, the heart, lungs, and muscles must all work together perfectly to allow you to run and jump. Physiology also helps us understand how our body reacts to different conditions outside. Whether the weather is very hot or very cold, your body has ways to help the inside of your body stay at just the right temperature. Physiology helps us understand how living creatures do all the things they do: eat, sleep, run, jump, even breathe and keep their heart beating! Physiology is the study of how animals and humans work.

Physiologists conduct their investigations at various levels of biological organisation, including: the molecular biology of cell components, single or groups of cells under culture, individual organs in isolation and the whole animal or human.

Physiology is a key life science essential to our understanding of human and animal function. In particular, with the increasing specialisation of biomedical research, physiology provides an essential framework within which new knowledge of cells and molecules can be applied to improving human health.

Physiologists play a major role in improving human health by actively searching for and elucidating the mechanisms of action of clinically-relevant drugs or treatments used for relieving the suffering of humans and animals. Physiology is the study of the mechanical, physical, and biochemical functions of living organisms. Physiology has traditionally been divided into plant physiology and animal physiology but the principles of physiology are universal, no matter what particular organism is being studied. For example, what is learned about the physiology of yeast cells can also apply to human cells.

The field of animal physiology extends the tools and methods of human physiology to non-human animal species. Plant physiology also borrows techniques from both fields. Its scope of subjects is at least as diverse as the tree of life itself. Due to this diversity of subjects, research in animal physiology tends to concentrate on understanding how physiological traits changed throughout the evolutionary history of animals.

Other major branches of scientific study that have grown out of physiology research include biochemistry, biophysics, paleobiology, biomechanics, and pharmacology.

GENETICS

When organisms reproduce the offspring tend to resemble their parents. They are not, however, identical to either parent, and they are not simply a mixture of the two parents.

For instance, in humans one parent is male and the other female, and their children are either male or female, not a mixture of the two traits. On the other hand a child might have her father's eye colour, and her mother's hair colour.

Genetics is concerned with explaining the behaviour of such inherited characteristics, in terms of the underlying genetic machinery which turns a single cell (the fertilised egg) into a worm, a carrot, or a human.

Genetics is the science of genes, heredity, and the variation of organisms. The word "genetics" was first suggested to describe the study of inheritance and the science of variation by the prominent British scientist William Bateson in a personal letter to Adam Sedgwick, dated April 18, 1905.

Bateson first used the term "genetics" publicly at the Third International Conference on Plant Hybridization (London, England) in 1906.

Heredity and variations form the basis of genetics. Humans applied knowledge of genetics in prehistory with the domestication and breeding of plants and animals. In modern research, genetics provides important tools for the investigation of the function of a particular gene, e.g., analysis of genetic interactions. Within organisms, genetic information generally is carried in chromosomes, where it is represented in the chemical structure of particular DNA (deoxyribonucleic acid) molecules.

Genes encode the information necessary for synthesizing the amino-acid sequences in proteins, which in turn play a large role in determining the final phenotype, or physical appearance, of the organism. In diploid organisms, a dominant allele on one chromosome will mask the expression of a recessive gene on the other. The only way that a recessive allele can be seen as a trait is if both parents contained that same recessive allele. Codominance is when both alleles for the same gene are considered dominant, in which case, both alleles will be present as traits.

The phrase to code for is often used to mean a gene contains the instructions about how to build a particular protein, as in the gene codes for the protein. The "one gene, one protein" concept is now known to be simplistic. For example, a single gene may produce multiple products, depending on how its transcription is regulated. Genes code for the nucleotide sequences in mRNA, tRNA and rRNA, required for protein synthesis.

Genetics determines much (but not all) of the appearance of organisms, including humans, and possibly how they act. Environmental differences and random factors also play a part. Monozygotic ("identical") twins, a clone resulting from the early splitting of an embryo, have the same DNA, but different personalities and fingerprints. Genetically-identical plants grown in colder climates incorporate shorter and less-saturated fatty acids to avoid stiffness. So genetics explains how like begets like. It also explains how, over longer time scales, living things change, or evolve, to produce the dazzling diversity of life.

WHY IS IT IMPORTANT?

Genetics allows us to understand normal events such as development, growth and ageing in terms of the underlying molecular machinery that direct these processes, and provides an understanding of why these processes go wrong in disease.

Genetics also provides us with tools to produce improved crops and livestock (we've been doing this for thousands of years

MOLECULAR BIOLOGY

Molecular biology is the study of biology at a molecular level. The field overlaps with other areas of biology and chemistry, particularly genetics and biochemistry. Molecular biology chiefly concerns itself with understanding the interactions between the various systems of a cell, including the interrelationship of DNA, RNA and protein synthesis and learning how these interactions are regulated.

Writing in Nature, William Astbury described molecular biology as:

"... not so much a technique as an approach, an approach from the viewpoint of the so-called basic sciences with the leading idea of searching below the large-scale manifestations of classical biology for the corresponding molecular plan. It is concerned particularly with the forms of biological molecules and ..... is predominantly three-dimensional and structural - which does not mean, however, that it is merely a refinement of morphology - it must at the same time inquire into genesis and function."

Molecular biology is the study of molecular underpinnings of the process of replication, transcription and translation of the genetic material. The central dogma of molecular biology where genetic material is transcribed into RNA and then translated into protein, despite being an oversimplified picture of molecular biology, still provides a good starting point for understanding the field. This picture, however, is undergoing revision in light of emerging novel roles for RNA.

Much of the work in molecular biology is quantitative, and recently much work has been done at the interface of molecular biology and computer science in bioinformatics and computational biology. As of the early 2000s, the study of gene structure and function, molecular genetics, has been amongst the most prominent sub-field of molecular biology.

Increasingly many other fields of biology focus on molecules, either directly studying their interactions in their own right such as in cell biology and developmental biology, or indirectly, where the techniques of molecular biology are used to infer historical attributes of populations or species, as in fields in evolutionary biology such as population genetics and phylogenetic. There is also a long tradition of studying biomolecules "from the ground up" in biophysics.

Techniques of Molecular Biology

Since the late 1950s and early 1960s, molecular biologists have learned to characterize, isolate, and manipulate the molecular components of cells and organisms. These components include DNA, the repository of genetic information; RNA, a close relative of DNA whose functions range from serving as a temporary working copy of DNA to actual structural and enzymatic functions as well as a functional and structural part of the translational apparatus; and proteins, the major structural and enzymatic type of molecule in cells.

Expression Cloning

One of the most basic techniques of molecular biology to study protein function is expression cloning. In this technique, DNA coding for a protein of interest is cloned (using PCR and/ or restriction enzymes) into a plasmid (known as an expression vector). This plasmid may have special promoter elements to drive production of the protein of interest, and may also have antibiotic resistance markers to help follow the plasmid.

This plasmid can be inserted into either bacterial or animal cells. Introducing DNA into bacterial cells is called transformation, and can be completed with several methods, including electroporation, microinjection, passive uptake and conjugation. Introducing DNA into eukaryotic cells, such as animal cells, is called transfection. Several different transfection techniques are available, including calcium phosphate transfection, liposome transfection, and proprietary transfection reagents such as Fugene. DNA can also be introduced into cells using viruses or pathenogenic bacteria as carriers. In such cases, the technique is called viral/bacterial transduction, and the cells are said to be transduced.

In either case, DNA coding for a protein of interest is now inside a cell, and the protein can now be expressed. A variety of systems, such as inducible promoters and specific cell-signaling factors, are available to help express the protein of interest at high levels. Large quantities of a protein can then be extracted from the bacterial or eukaryotic cell. The protein can be tested for enzymatic activity under a variety of situations, the protein may be crystallized so its tertiary structure can be studied, or, in the pharmaceutical industry, the activity of new drugs against the protein can be studied.

MORPHOLOGY

The term morphology in biology refers to the outward appearance (shape, structure, colour, pattern) of an organism or taxon and its component parts.

Also in use is the term "gross morphology", which refers to the prominent or principal aspects of an organism or taxon's morphology. A description of an organism's gross morphology would include, for example, its overall shape, overall colour, main markings etc. but not finer details.

Most taxa differ morphologically from other taxa. Typically closely related taxa differ much less than more distantly related ones, but there are exceptions to this. Cryptic species are species which look very similar, or perhaps even outwardly identical, but are reproductively isolated. Conversely, sometimes unrelated taxa acquire similar appearance through convergent evolution.

SYSTEMATICS

All life on Earth is united by evolutionary history; we are all evolutionary cousins – twigs on the tree of life. Phylogenetic systematics is the formal name for the field within biology that reconstructs evolutionary history and studies the patterns of relationships among organisms. Unfortunately, history is not something we can see. It has only happened once and only leaves behind clues as to what happened. Systematists use these clues to try to reconstruct evolutionary history. Systematics is the study of the historical relationships of groups of biological organisms – the recognition and understanding of biodiversity. Systematics differs from ecology in that the latter is concerned with the interactions of taxa and individuals in a particular time, while the former is concerned only with the relationships of hierarchic lineages through time. Systematics includes the processes of identifying the basic systematic unit (the species), discovering the patterns of relationships of species at successively higher levels, building classifications based on these patterns and naming appropriate taxa (taxonomy), and the application of this pattern knowledge to studying changes in organismal features through time. It also includes the building and maintenance of biodiversity collections, upon which all the products of systematic studies are based. These are museum collections of preserved specimens of all kinds; such a museum collection of plant specimens is called a herbarium.

Systematics has undergone a revolution in its basic paradigm over the last 40 years. This revolution is just the latest step in a progression that has paralleled advances in other academic disciplines through the history of man. Some concept of relationship, the idea, for example, that a bluebird is more like an ostrich than it is like a an antelope, has existed since the early sentience of man. During the 1700's, very basic, utility-driven systems of classification (such as those used by the herbalists through the Middle Ages, and notably, by Linnaeus) began to be replaced by "natural" systems that were based on a comparison of large numbers of features, or characters, of the organisms under study. During the next century, the concept of evolution gave causal explanation for the patterns that were being observed — for how a group of jawbones in reptiles could be transformed into the ear bones of mammals, as an example.

A new classification criterion was then possible — that taxa be grouped according to evolutionary relationship. An intrinsic part of this idea is that groups of organisms change over time. Yet it took until the middle of the 20th century for biologists to realize that it is the later form of a character in time, the "advanced state", that gives us the best clue to phylogenetic relationships and that can be used to group organisms together because it signifies that they share a common history. This realization is the key component to the methodology known as cladistics, which is our current systematic paradigm. The method uses these advanced characters, or synapomorphies, to produce explicit, testable patterns of phylogenetic relationship among organisms. In recent years, researchers have continued to refine the methodology, seeking the best ways by which to analyze character data to produce these patterns, as well as devising methods for evaluating the strength of these hypotheses, developing new sources of character information, and realizing the power of the resulting patterns when applied to any questions that deal with the evolution of organisms or their characters.

The study of evolution is often considered to be closely related to systematics. In fact, the two are essentially cause and effect. Although systematics can be done without regard to any process, since in its starkest form it is only a study of patterns without regard to how they came about, most researchers see evolution as the causal agent for these patterns. Hence, studies of evolution examine the processes, at the individual and population level, that lead to the patterns that we study in systematics.

The Roles and Products of Systematics in Modern Biology?

As the sub-discipline of biology that investigates relationships of taxa, systematics is the foundation for comparative biology. Comparative biology is that type of study that attempts to relate features of one organism, or type of organism, to features in another type of organism. This always is a question of homology, or sameness due to common evolutionary origin. In systematic studies we hypothesize homology of features among taxa and then gather data to test these hypotheses. This is important because appearance alone is often not a good indicator that features in different taxa are homologous – many times similar structures will evolve independently in different lineages. If they are homologous, we expect that they will share many things because of their common ancestry, while if they are not, it is impossible to predict just how similar they will be. Hence, any study that asks why or how about a feature in more than one taxon, and draws comparative conclusions about them, rests on a systematic foundation.

You can identify specific roles for systematic studies and the patterns they produce, as follows:

1. *Systematists identify and document Earth's biodiversity, and organize this information in a form that can be utilized by others:* A long-standing role for systematists is that of going into the field and collecting samples of organisms, then comparing them with known specimens in order to determine whether something significantly different has been found, a new species. Such work depends upon the expertise of specialists

who are intimately familiar with the natural variation in a particular group. This expertise can only be gained by first-hand experience with the organisms, both in the field and in biodiversity collections. Once species have been defined, names are given to them according to rules of nomenclature for the group. Higher level taxa (genera, families, etc.), which are successively larger assemblages of species, can then be named based on the phylogenetic relationships of the species. The resulting classifications provide a basis for communication about taxa for the scientific community and for the world at large. Because biodiversity collections are intended to be permanent, and are assembled over time, they provide a way of documenting change in the world's flora and fauna, and can therefore provide supporting evidence for putative causal processes such as the "greenhouse effect."

2. *Systematics is the study of the history of life on Earth:* Once we know what organisms exist, we can then ask questions about how they came to be as they are today. Phylogenetic analysis allows us to combine data from extant organisms with data from fossils to provide hypotheses of relationship — to actually reconstruct the history of life. It allows us to determine which taxa are most closely related to the dinosaurs, which characters may have been key to the success of the flowering plants, and how many times HIV may have shifted hosts (e.g., between simians and man). This is because our phylogenetic hypotheses are both hypotheses of relationships of taxa and of character transformation. These patterns are framed as hypotheses because they are always subject to testing by additional characters.

3. *Phylogenetic patterns that result from systematic studies, and classifications derived from them, have predictive value:* Common ancestry means that organisms will

have a greater or lesser amount of expected similarity depending upon how closely related they are. This principle can be put to immediate use when one seeks other taxa that may possess a feature of interest found in a specific taxon. For example, the anti-cancer compound taxol was isolated from a particular species of conifer, the Pacific Yew (Taxus brevifolia). Where else would we look to find other sources of this compound? The logical place to look would be in taxa that are most closely related to T. brevifolia. Armed with information about relationships in the genus, researchers found taxol in a closely related species, the Euopean Yew (T. baccata). This alternate source is less costly and will alleviate pressure on the rarer T. brevifolia. There is no guarantee in cases such as this that we will find what we are looking for, since the substance may have arisen only within one species, but rather than searching blindly, we increase our chances of success by looking in related species. Having the systematic guide for where to look is especially important in large groups (a genus of say, 500 species) to maximize use of time and resources. The list of biodiversity attributes of interest to man that such information can be applied to is endless, including all types of substances and qualities derived from living sources.

4. *Systematics provides a basis for biodiversity conservation priorities:* With increasing pressures from a growing world population and resulting pressure on biotic resources, we now and in the future have to make difficult decisions about what parts of the Earth will be maintained in a "natural" state in order to conserve the biodiversity present there. How do we decide, given limited resources, which to protect? If we decide that we want to maximize biodiversity, then the phylogenetic patterns produced by systematists give us a way to prioritize areas based

upon the diversity they contain. In order to maximize diversity, it makes sense to try to preserve groups from throughout the tree of life, rather than large numbers from one branch. In this way we will tend to preserve a wider array of features that have potential use for humans, though their uses may presently be unknown, but it does mean that we have to know something about the relationships of the organisms involved.

5. *Systematics provides independent evidence for patterns of geological change:* The continents have not always held the positions on the Earth that they do today, nor have they been the same size and shape. Geologists use data from the Earth itself to reconstruct past arrangements of land masses. However, there is an independent source of data for such reconstructions, which lies in the current distribution of taxa when viewed in the light of their relationships. When continents fragment, the taxa that live on them record this change, since the separated taxa will then have their own history, but share the common ancestor that was once continuously distributed. By constructing organismal phylogenies and mapping on current distributions of taxa, and doing this for many groups, general patterns emerge that may best be explained by historical geological events.

6. *Systematists and systematic collections provide identification services and documentation of identity:* Another crucial role for systematists is that of identification specialists. They are in a unique position to provide this service, with experience and the necessary tools. The importance of correct identification cannot be overstated — when a life, for instance, hangs in the balance depending on whether the plant or mushroom that has been ingested is poisonous or not, this service is critical. Other types of biological research are essentially valueless if their

subjects are misidentified, since closely related taxa can have very different properties and generalizations must be made carefully. Hence, documentation is important so that subsequent investigators can confirm indentifications. The only lasting way to document indentity is to deposit a voucher specimen in an appropriate collection. Studies that do not utilize this service will have less value in the long term because of the impossibility of verifying indentification.

ECOLOGY

Ecology is the study of "what lives where and why". It is about where animals and plants are found and how they interact with each other and their environment. Their environment has biological features - other plants and animals, including people. It also has physical features, such as temperature and rainfall. Animals and plants have very complex relationships with their surroundings.

Ecology is generally spoken of as a new science, having only become prominent in the second half of the 20th Century. Nonetheless, ecological thinking at some level has been around for a long time, and the principles of ecology have developed gradually, closely intertwined with the development of other biological disciplines. Thus, one of the first ecologists may have been Aristotle or perhaps his student, Theophrastus, both of whom had interest in many species of animals. Theophrastus described interrelationships between animals and between animals and their environment as early as the 4th century BC.